Icons of Mathematics

An Exploration of
Twenty Key Images

The Dolciani Mathematical Expositions

NUMBER FORTY-FIVE

Icons of Mathematics

An Exploration of Twenty Key Images

Claudi Alsina
Universitat Politècnica de Catalunya

Roger B. Nelsen
Lewis & Clark College

Published and Distributed by
The Mathematical Association of America

10 0669859 8

The DOLCIANI MATHEMATICAL EXPOSITIONS series of the Mathematical Association of America was established through a generous gift to the Association from Mary P. Dolciani, Professor of Mathematics at Hunter College of the City University of New York. In making the gift, Professor Dolciani, herself an exceptionally talented and successful expositor of mathematics, had the purpose of furthering the ideal of excellence in mathematical exposition.

The Association, for its part, was delighted to accept the gracious gesture initiating the revolving fund for this series from one who has served the Association with distinction, both as a member of the Committee on Publications and as a member of the Board of Governors. It was with genuine pleasure that the Board chose to name the series in her honor.

The books in the series are selected for their lucid expository style and stimulating mathematical content. Typically, they contain an ample supply of exercises, many with accompanying solutions. They are intended to be sufficiently elementary for the undergraduate and even the mathematically inclined high-school student to understand and enjoy, but also to be interesting and sometimes challenging to the more advanced mathematician.

1. *Mathematical Gems*, Ross Honsberger

2. *Mathematical Gems II*, Ross Honsberger

3. *Mathematical Morsels*, Ross Honsberger

4. *Mathematical Plums*, Ross Honsberger (ed.)

5. *Great Moments in Mathematics (Before 1650)*, Howard Eves

6. *Maxima and Minima without Calculus*, Ivan Niven

7. *Great Moments in Mathematics (After 1650)*, Howard Eves

8. *Map Coloring, Polyhedra, and the Four-Color Problem*, David Barnette

9. *Mathematical Gems III*, Ross Honsberger

10. *More Mathematical Morsels*, Ross Honsberger

11. *Old and New Unsolved Problems in Plane Geometry and Number Theory*, Victor Klee and Stan Wagon

12. *Problems for Mathematicians, Young and Old*, Paul R. Halmos

13. *Excursions in Calculus: An Interplay of the Continuous and the Discrete*, Robert M. Young

14. *The Wohascum County Problem Book*, George T. Gilbert, Mark Krusemeyer, and Loren C. Larson

15. *Lion Hunting and Other Mathematical Pursuits: A Collection of Mathematics, Verse, and Stories by Ralph P. Boas, Jr.*, edited by Gerald L. Alexanderson and Dale H. Mugler

16. *Linear Algebra Problem Book*, Paul R. Halmos

MAA Service Center
P.O. Box 91112
Washington, DC 20090-1112
1-800-331-1MAA   FAX: 1-301-206-9789

*Dedicated to the icons of our lives —
our heroes, our mentors, and our loved ones.*

Preface

*Of all of our inventions for mass communication, pictures
still speak the most universally understood language.*

Walt Disney

An *icon* (from the Greek εικών, "image") is defined as "a picture that is
universally recognized to be representative of something." The world is full
of distinctive icons. Flags and shields represent countries, graphic designs
represent commercial enterprises; paintings, photographs and even people
themselves may evoke concepts, beliefs and epochs. Computer icons are essential tools for working with a great variety of electronic devises.

What are the icons of mathematics? Numerals? Symbols? Equations?
After many years working with visual proofs (also called "proofs without
words"), we believe that certain geometric diagrams play a crucial role in
visualizing mathematical proofs. In this book we present twenty of them,
which we call icons of mathematics, and explore the mathematics that lies
within and that can be created. All of our icons are two-dimensional; three-
dimensional icons will appear in a subsequent work.

Some of the icons have a long history both inside and outside of mathematics (yin and yang, star polygons, the Venn diagram, etc.). But most of
them are essential geometrical figures that enable us to explore an extraor-
dinary range of mathematical results (the bride's chair, the semicircle, the
rectangular hyperbola, etc.).

Icons of Mathematics is organized as follows. After the Preface we present
a table with our twenty key icons. We then devote a chapter to each, illustrating its presence in real life, its primary mathematical characteristics and
how it plays a central role in visual proofs of a wide range of mathematical facts. Among these are classical results from plane geometry, properties
of the integers, means and inequalities, trigonometric identities, theorems
from calculus, and puzzles from recreational mathematics. As the American

actor Robert Stack once said (speaking of icons of a different sort), "these are icons to be treasured."

Each chapter concludes with a selection of Challenges for the reader to explore further properties and applications of the icon. After the chapters we give solutions to all the Challenges in the book. We hope that many readers will find solutions that are superior to ours. *Icons of Mathematics* concludes with references and a complete index.

As with our previous books with the MAA, we hope that both secondary school and college and university teachers may wish to use portions of it as a supplement in problem solving sessions, as enrichment material in a course on proofs and mathematical reasoning, or in a mathematics course for liberal arts students.

Special thanks to Rosa Navarro for her superb work in the preparation of preliminary drafts of this manuscript. Thanks too to Underwood Dudley and the members of the editorial board of the Dolciani series for their careful reading of an earlier draft of the book and their many helpful suggestions. We would also like to thank Carol Baxter, Beverly Ruedi, and Rebecca Elmo of the MAA's book publication staff for their expertise in preparing this book for publication. Finally, special thanks to Don Albers, the MAA's editorial director for books, who as on previous occasions encouraged us to pursue this project and guided its final production.

Claudi Alsina
Universitat Politècnica de Catalunya
Barcelona, Spain

Roger B. Nelsen
Lewis & Clark College
Portland, Oregon

Twenty Key Icons
of Mathematics

The Bride's Chair	Zhou Bi Suan Jing	Garfield's Trapezoid	The Semicircle
Similar Figures	Cevians	The Right Triangle	Napoleon's Triangles
Arcs and Angles	Polygons with Circles	Two Circles	Venn Diagrams
Overlapping Figures	Yin and Yang	Polygonal Lines	Star Polygons
Self-similar Figures	Tatami	The Rectangular Hyperbola	Tiling

Contents

CHAPTER **1**

The Bride's Chair

> *... it has been said that the art of geometry is to reason well from false diagrams.*
>
> Jean Dieudonné
> *Mathematics—The Music of Reason*

Perhaps the best-known theorem in mathematics is the *Pythagorean theorem*, Proposition 47 in Book I of the *Elements* of Euclid (circa 300 BCE), which states: *In right-angled triangles the square on the side opposite the right angle equals the sum of the squares on the sides containing the right angle.* Perhaps one of the most recognizable images in mathematics is the figure that often accompanies the Pythagorean theorem, an icon variously known as the *bride's chair*, the *peacock's tail*, the *windmill*, and the *Franciscan's cowl*. In Figure 1.1 we see the bride's chair in the book *Los Seis Libros Primeros de la Geometria de Euclides*, a Spanish translation of the *Elements* published in Seville in 1576, and on a Greek postage stamp from 1955.

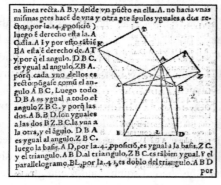

Figure 1.1.

1

Since the bride's chair figures in many proofs of the Pythagorean theorem, we begin by presenting several proofs that employ it in different ways. We then generalize to form the *Vecten configuration* (squares on the sides of a general triangle), and present several surprising results whose proofs use it. All of the results we present for the Vecten configuration also hold for the bride's chair.

Why the name *the bride's chair*?

Florian Cajori [Cajori, 1899] answers as follows: "Some Arabic writers, Behâ Eddîn for instance, call the Pythagorean theorem 'figure of the bride.' Curiously enough, this romantic appellation appears to have originated from a mistranslation of the Greek word νυμφη, applied to the theorem by a Byzantine writer of the 13th century. This Greek word admits two meanings, 'bride' and 'winged insect.' The figure of the right triangle with the three squares suggests an insect, but Behâ Eddîn apparently translated the word as 'bride.'"

1.1   The Pythagorean theorem—Euclid's proof and more

> *By any measure, the Pythagorean theorem is the most famous statement in all of mathematics.*
>
> Eli Maor
> *The Pythagorean Theorem: A 4,000-Year History*

There may well be more proofs of the Pythagorean theorem than any other theorem in mathematics. The classic book *The Pythagorean Proposition* by Elisha Scott Loomis [Loomis, 1968] has over 370 proofs, and Alexander Bogomolny's website www.cut-the-knot.org has 92 proofs, some interactive. Many of them employ the bride's chair.

In his proof of the Pythagorean theorem, Euclid first shows that the two overlapping triangles in Figure 1.2a are congruent and thus have the same area, from which he concludes that the areas of the shaded square and rectangle in Figure 1.2b are equal. He similarly reaches the same conclusion for the triangles in Figure 1.2c and the shaded square and rectangle in Figure 1.2d, which establishes the theorem.

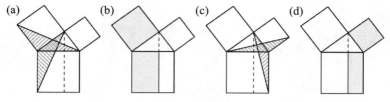

Figure 1.2.

In Figure 1.3 we see a modern dynamic version [Eves, 1980] of Euclid's proof, where area-preserving transformations change squares into parallelograms and rectangles to establish the theorem.

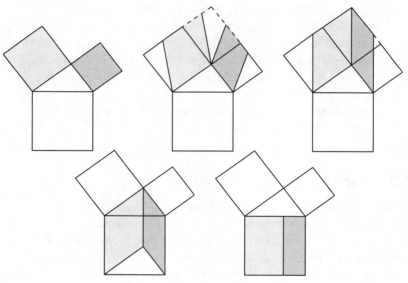

Figure 1.3.

In Figure 1.4 the bride's chair is used in three dissection proofs of the theorem, where the squares on the sides of the triangle adjacent to the right angle are cut into pieces that are reassembled to form the square on the hypotenuse. Although there is some question about their proper attribution, the first is often attributed to Liu Hui (3rd century AD) [Wagner, 1985], the second to Thābit ibn Qurra (836–901) and the third to Henry Perigal (1801–1899) [Frederickson, 1997]. Other dissection proofs are possible: see Section 20.4. In Chapters 2, 3, and 18 we use other icons to present additional proofs.

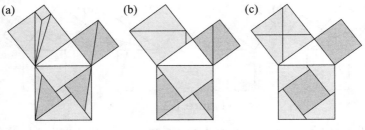

Figure 1.4.

The bride's chair in three dimensions

What is the three-dimensional analog of the Pythagorean theorem? The most common answer is the relationship between the diagonal d of a box with edges a, b, and c: $d^2 = a^2 + b^2 + c^2$. However, the French mathematician Jean Paul de Gua de Malves (1713–1785) considered the *right tetrahedron* (a tetrahedron with three faces perpendicular to one another at one vertex) and proved that the square of the area of the face opposite the vertex where the three mutually perpendicular faces meet is equal to the sum of the squares of the areas of the other three faces.

1.2   The Vecten configuration

Replacing the right triangle in the bride's chair by an arbitrary triangle yields *Vecten's configuration*, as illustrated in Figure 1.5a. Little is known about Vecten (not even his first name) except that he was "professeur de mathématiques spéciales" at the Lycée de Nîmes in France during 1810–1816. He is remembered in France for being one of the first to study the "windmill," his name for what we now call the Vecten configuration [Ayme, 2010]. He published 22 papers on it in the journal *Annales* of his colleague Joseph Diaz Gergonne (1771–1859).

Connecting adjacent vertices of the squares in the Vecten configuration yields three more triangles, called the *flanks* of the configuration, the gray triangles illustrated in Figure 1.5b.

In a Vecten configuration each of the flanks has the same area as the original triangle. To prove this [Snover, 2000] we erase the squares and rotate the flanks 90° counterclockwise as shown in Figure 1.6a into the positions in Figure 1.6b, where each flank has the same base length and altitude (and hence the same area) as the original triangle.

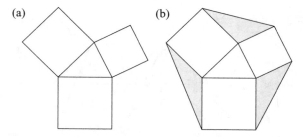

Figure 1.5.

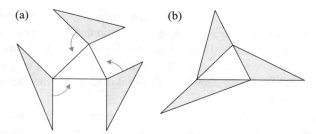

Figure 1.6.

As a corollary, if two squares share a vertex and form two flank triangles, as shown in Figure 1.7a, then the triangles have the same area. This follows immediately from the Vecten configuration in Figure 1.5b. Furthermore, the centers of the squares along with the midpoints of the sides of the flanks that are not sides of squares are the vertices of another square, a result known as the *Finsler-Hadwiger theorem*. See Figure 1.7b. For a proof, see Challenge 1.6.

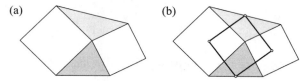

Figure 1.7.

There are many remarkable properties of the Vecten configuration; we mention but a few. In Figure 1.8a we observe that the medians of the flanks coincide with the altitudes of the central triangle. In addition each median is one-half the length of the opposite side of the central triangle, i.e., $2|AP| = |BC|$, $2|BQ| = |AC|$, and $2|CR| = |AB|$.

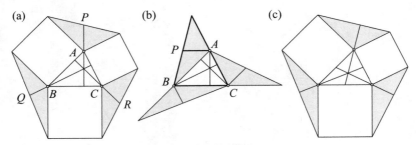

Figure 1.8.

To prove this [Warburton, 1996], we again rotate the flanks 90° counter-clockwise, as shown in Figure 1.8b. After the rotation, AP is parallel to BC and $2|AP| = |BC|$. Thus in Figure 1.8a, AP is perpendicular to BC, and similarly for the other medians. Finally, since the central triangle is a flank for each of its flanks, the medians of the central triangle are the altitudes of the flanks, as shown in Figure 1.8c.

In Figure 1.9 we let P_a, P_b, and P_c denote the centers of the squares in a Vecten configuration, and draw a line segment connecting two of them (say P_a and P_b) and a line segment connecting the third to the opposite vertex of the triangle (here P_c to C). We now prove that the two line segments $P_a P_b$ and CP_c are perpendicular and equal in length. Our proof is from [Coxeter and Greitzer, 1967].

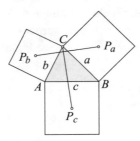

Figure 1.9.

Draw triangles ABK and CBK as illustrated in Figure 1.10a and reduce each in size by a factor of $\sqrt{2}/2$, as illustrated in Figure 1.10b. The images of the segment BK are parallel and equal in length. Rotate the light gray triangle 45° clockwise to ACP_c and the dark gray triangle 45° counter-clockwise to $CP_a P_b$, as shown in Figure 1.10c. As a result, $P_a P_b$ and CP_c are perpendicular and equal in length.

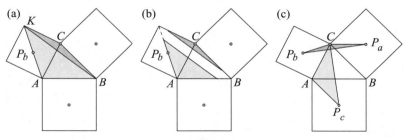

Figure 1.10.

The three line segments AP_a, BP_b, and CP_c (see Figure 1.11) are therefore perpendicular to $P_b P_c$, $P_a P_c$, and $P_a P_b$, respectively. Since they contain the altitudes of triangle $P_a P_b P_c$ they are concurrent. Their point of intersection is known as the *Vecten point* of the triangle ABC.

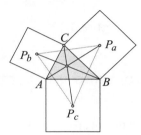

Figure 1.11.

1.3 The law of cosines

We now present a nice proof of the law of cosines using both the bride's chair and the Vecten configuration, derived from [Sipka, 1988]. Given an arbitrary triangle with sides a, b, c, construct the Vecten configuration as illustrated in Figure 1.12a, and draw the altitude to side BC. Then rotate the line segment counterclockwise about vertex A as shown. The length of the altitude is $b \sin C$, so we draw squares with sides $b \sin C$ and $a - b \cos C$. The two squares and the one with area c^2 now form a bride's chair, as shown in Figure 1.12b, so by the Pythagorean theorem we have

$$
\begin{aligned}
c^2 &= (b \sin C)^2 + (a - b \cos C)^2 \\
&= a^2 + b^2 - 2ab \cos C,
\end{aligned}
$$

as desired (and similarly $a^2 = b^2 + c^2 - 2bc \cos A$ and $b^2 = a^2 + c^2 - 2ac \cos B$).

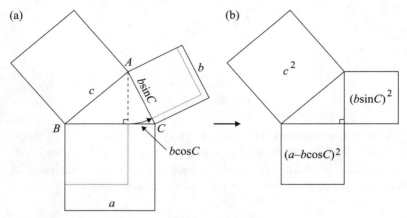

Figure 1.12.

1.4   Grebe's theorem and van Lamoen's extension

Extend the outermost sides of the squares in Vecten's configuration until they intersect to form a triangle, as illustrated in Figure 1.13a. The large triangle is clearly similar to the central triangle, as its sides are parallel to those of the central triangle. But more that that—the two triangles are *homothetic*, i.e., lines joining corresponding points (here the vertices) of the two triangles are concurrent, a result known as *Grebe's theorem*. This point of concurrency (or *point of homothety*) is sometimes called the *Lemoine point* or *Grebe's point* of the triangle.

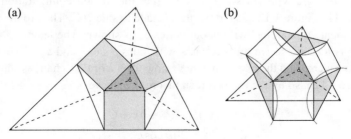

Figure 1.13.

Similarly, consider the triangle formed by the circumcenters of the flanks, as illustrated in Figure 1.13b. Since the circumcenter of each flank lies at the intersection of the perpendicular bisectors of its sides, the resulting triangle has as its sides the lines bisecting the squares and parallel to the sides of the central triangle in the Vecten configuration. Hence it is similar to the central triangle, and, as shown by F. van Lamoen [van Lamoen, 2001], homothetic to the central triangle with the same point of homothety as in Grebe's theorem.

1.5 Pythagoras and Vecten in recreational mathematics

Sam Loyd (1841–1911) was one of the best-known creators of mathematical puzzles of his time. The book *Sam Loyd's Cyclopedia of 5000 Puzzles, Tricks, and Conundrums (With Answers)* [Loyd, 1914] contains about 2700 mathematical puzzles and recreations.

In "Pythagoras' Classical Problem," from page 101 of the *Cyclopedia,* Loyd says: "Take a piece of paper of the dimensions of the two squares, as shown in the picture, and cut it into three pieces which will fit together and make a perfect square." See Figure 1.14 for a picture of Loyd and this problem.

The name of the puzzle indicates the solution—the Pythagorean theorem, specifically the dissection proof in Figure 1.4b. See Figure 1.15.

The Vecten configuration appears in "The Lake Puzzle," from page 267 of the *Cyclopedia,* which concerns a triangular lake surrounded by three square plots of land (see Figure 1.16). Loyd writes: "The question which I ask our

Figure 1.14.

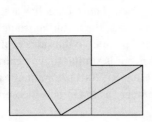
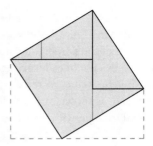

Figure 1.15.

puzzlists who revel in just such questions, is to determine just how many acres there would be in that triangular lake, surrounded as shown by three square plots of 370, 116, and 74 acres."

Figure 1.16.

Let a, b, c denote the sides, K the area, and $s = (a + b + c)/2$ the semiperimeter of the triangle. Heron's formula $K = \sqrt{s(s-a)(s-b)(s-c)}$ is equivalent to $16K^2 = 2(a^2b^2 + b^2c^2 + c^2a^2) - (a^4 + b^4 + c^4)$, which expresses the square of the area of a triangle in terms of the areas of the squares on the sides. With $a^2 = 370, b^2 = 116$, and $c^2 = 74$ we have $16K^2 = 1936$, and thus $K = 11$ acres.

For an alternate solution, if we note that $370 = 9^2 + 17^2$, $116 = 4^2 + 10^2$, and $74 = 5^2 + 7^2$, then we have the situation in Figure 1.17 from which it follows that the area K of the lake (the gray triangle) is

$$K = \frac{9 \cdot 17}{2} - \left(4 \cdot 7 + \frac{4 \cdot 10}{2} + \frac{5 \cdot 7}{2}\right) = 11 \text{ acres.}$$

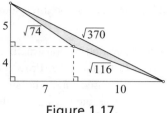

Figure 1.17.

See Challenge 20.5 for another puzzle concerning the Vecten configuration.

Imagining Pythagoras

Portraits and busts of mathematicians from antiquity arise from the imaginations of artists and sculptors. The fame of the Pythagorean theorem through the centuries has motivated a great collection of images of its namesake. From left to right in Figure 1.18 we see a bust from the Capitoline Museums in Rome, an illustration from the *Nuremburg Chronicle* (1493), detail from Rafael's *The School of Athens* (1509), and a postage stamp issued by San Marino in 1982.

Figure 1.18.

1.6 Challenges

1.1. The right triangle in the postage stamp in Figure 1.1 is the 3-4-5 right triangle. (a) Are there other right triangles whose sides are in arithmetic progression? (b) Are there bride's chairs in which the areas of the three squares are in arithmetic progression? (c) Are there bride's chairs in which the areas of the three squares are in geometric progression?

1.2. Enlarge the Vecten configuration by constructing squares on the sides of the flanks, as shown in Figure 1.19. Prove (a) the sum of the areas of the outer squares is three times the sum of the areas of the inner squares, i.e., if x, y, z denote the sides of the outer squares, then $x^2 + y^2 + z^2 = 3(a^2 + b^2 + c^2)$ and (b) if the original configuration is a bride's chair with $a^2 + b^2 = c^2$ (as in Figure 1.19b), then $x^2 + y^2 = 5z^2$.

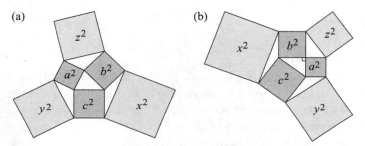

Figure 1.19.

1.3. In the enlarged Vecten configuration of Figure 1.19a, let P_a, P_b, P_c, P_x, P_y, P_z denote the centers of the squares with sides a, b, c, x, y, z, respectively. Show that vertex A is the midpoint of $P_a P_x$, vertex B is the midpoint of $P_b P_y$, and vertex C is the midpoint of $P_c P_z$. See Figure 1.20.

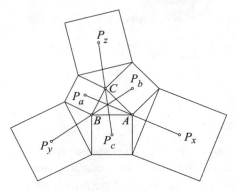

Figure 1.20.

1.4. *Sangaku* are Japanese geometry theorems that were often written on wooden tablets during the Edo period (1603–1867) and hung on Buddhist temples as offerings. This problem is from 1844 in the Aichi prefecture [Konhauser et al., 1996]: Five squares are arranged as shown in Figure 1.21. Show that the area of the shaded triangle is equal to the area of the shaded square.

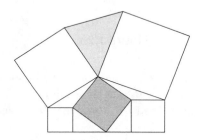

Figure 1.21.

1.5. Consider the bride's chair without the square on the hypotenuse, as illustrated in Figure 1.22. Draw line segments from each acute angle

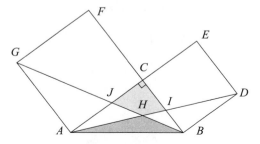

Figure 1.22.

vertex to a vertex of the square on the opposite side as shown. Which has the larger area, triangle ABH or quadrilateral $HIJC$?

1.6. Prove the *Finsler-Hadwiger theorem*: Let the squares $ABCD$ and $AB'C'D'$ share a common vertex A, as shown in Figure 1.23. Then the midpoints Q and S of the segments BD' and $B'D$ together with the centers R and T of the original squares form another square $QRST$.

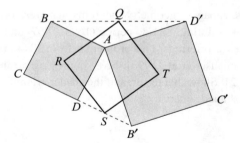

Figure 1.23.

1.7. Squares $ACED$ and $BCFG$ are constructed outwardly on the sides of an arbitrary triangle ABC, as shown in Figure 1.24. Let P denote the intersection of AF and BE. Prove that D, P, and G are collinear.

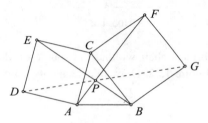

Figure 1.24.

CHAPTER 2

Zhou Bi Suan Jing

The hsuan-thu *diagram of the* Zhou bi suan jing *represents the oldest recorded proof of the "Pythagorean" theorem.*

Frank J. Swetz and T. I. Kao
Was Pythagoras Chinese?

The *Zhou bi suan jing* (周髀算经), "The Arithmetical Classic of the Gnomon and Circular Paths of Heaven," is a Chinese text dating from the Zhou dynasty (1046–256 BCE). Although primarily an astronomy text, it also discusses right triangle geometry. The image in Figure 2.1, called the *hsuan-thu* in Chinese, appears in this text; however, we will call this icon the *Zhou bi suan jing*, the title of the text in which it appears.

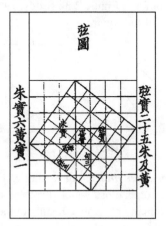

Figure 2.1.

We begin by examining the *Zhou bi suan jing* proof of the Pythagorean theorem, and then generalize the icon to a rectangular form to prove results

15

such as the arithmetic mean-geometric mean inequality, the Cauchy-Schwarz inequality, and two trigonometric formulas.

2.1 The Pythagorean theorem—a proof from ancient China

While the *Zhou bi suan jing* illustrates the Pythagorean theorem for a 3-4-5 right triangle, it is easily modified to prove the theorem in general. For a right triangle with legs a and b, hypotenuse c, and area T, we express the area of a square with sides $a + b$ in two different ways: first as $4T + c^2$, and then as $4T + a^2 + b^2$, hence $a^2 + b^2 = c^2$. See Figure 2.2.

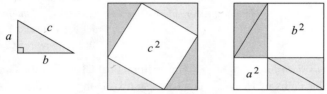

Figure 2.2.

The *Zhou bi suan jing* can also be used to prove that: *The internal bisector of the right angle of a right triangle bisects the square on the hypotenuse.* See Figure 2.3a. For a proof, see Figure 2.3b [Eddy, 1991]. The bisector also passes through the center of the square.

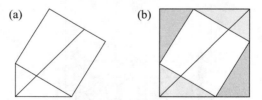

Figure 2.3.

Bisecting the *Zhou bi suan jing* along a diagonal of the inner square produces two congruent trapezoids, which leads to another proof of the Pythagorean theorem and many other nice results in the next chapter.

The hundred greatest theorems

In July 1999 Paul and Jack Abad presented a list of "The Hundred Greatest Theorems." Criteria were used for the ranking were the place the theorem holds in the literature, the quality of the proof, and the unexpectedness of the result. The Pythagorean theorem is number four on the list. The number one ranked theorem is also one credited to the school of Pythagoras, the irrationality of the square root of two (see Section 13.2).

In 1971 Nicaragua issued a series of ten postage stamps entitled *The Ten Mathematical Formulas That Changed the Face of the Earth*. The fifth stamp in the series is shown in Figure 2.4.

Figure 2.4.

2.2 Two classical inequalities

Versions of the *Zhou bi suan jing* employing a rectangle with four triangles leads to visual proofs of two classical inequalities—the *arithmetic mean-geometric mean inequality* (or *AM-GM inequality*) for two numbers and the two-dimensional version of the *Cauchy-Schwarz inequality*. In these proofs, we use the fact that the area of a parallelogram with sides a and b and a vertex angle θ is $ab \sin \theta$ and the inequality $\sin \theta \leq 1$.

The arithmetic and geometric means of a and b are $(a + b)/2$ and $\sqrt{ab}$, respectively, and the AM-GM inequality states that for a and b positive,

$$\sqrt{ab} \leq (a + b)/2. \tag{2.1}$$

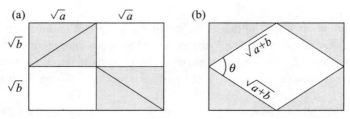

Figure 2.5.

Each white rectangle in Figure 2.5a has area $\sqrt{ab}$, the white paral-lelogram in Figure 2.5b has area $(\sqrt{a+b})^2 \sin\theta$, and hence $2\sqrt{ab} = (\sqrt{a+b})^2 \sin\theta \le a + b$, which establishes the inequality.

For real numbers a, b, x, y, the Cauchy-Schwarz inequality states that

$$|ax + by| \le \sqrt{a^2 + b^2}\sqrt{x^2 + y^2}. \tag{2.2}$$

Since $|ax + by| \le |a||x| + |b||y|$, it suffices to show that $|a||x| + |b||y| \le \sqrt{a^2 + b^2}\sqrt{x^2 + y^2}$, which we do in Figure 2.6 [Kung, 2008]. The sum of the areas of the white rectangles in Figure 2.6a is $|a||x| + |b||y|$, the area of the parallelogram in Figure 2.6b is $\sqrt{a^2 + b^2}\sqrt{x^2 + y^2}\sin\theta$, and hence

$$|ax + by| \le |a||x| + |b||y|$$
$$= \sqrt{a^2 + b^2}\sqrt{x^2 + y^2}\sin\theta \le \sqrt{a^2 + b^2}\sqrt{x^2 + y^2}.$$

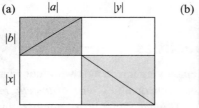
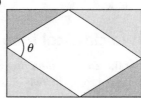

Figure 2.6.

Additional proofs using different icons appear in Chapters 13 and 18.

2.3 Two trigonometric formulas

The reader may have noticed that the *Zhou bi suan jing* is useful for illus-trating expressions that involve a sum of two products of two numbers, i.e., expressions of the form $pq + rs$. Two trigonometric formulas in this form are the formulas for the sine of the sum and the cosine of the difference of two angles or numbers:

$$\sin(\alpha + \beta) = \sin\alpha\cos\beta + \cos\alpha\sin\beta,$$
$$\cos(\alpha - \beta) = \cos\alpha\cos\beta + \sin\alpha\sin\beta.$$

In Figure 2.7a, the area of the white parallelogram is $\sin(\alpha + \beta)$, and is clearly equal to the sum of the areas of the two white rectangles in Figure 2.7b [Priebe and Ramos, 2000].

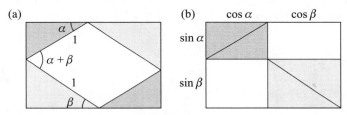

Figure 2.7.

The formula for the cosine of a difference is proven similarly, see Challenge 2.2. In the next chapter we present alternative proofs of these identities, as well as proofs of other addition and subtraction formulas for trigonometric functions.

2.4 Challenges

2.1. When does equality hold in the AM-GM inequality (2.1) and the Cauchy-Schwarz inequality (2.2)? (Hint: The answers lie in Figures 2.5 and 2.6).

2.2. Illustrate $\cos(\alpha - \beta) = \cos\alpha\cos\beta + \sin\alpha\sin\beta$ using a rectangular version of the *Zhou bi suan jing*.

2.3. Illustrate $|a\sin t + b\cos t| \le \sqrt{a^2 + b^2}$ for real a, b, t using a rectangular version of the *Zhou bi suan jing*.

2.4. For real numbers a, b, c, x, y, z, show that

$$|ax + by + cz| \le \sqrt{a^2 + b^2 + c^2}\sqrt{x^2 + y^2 + z^2}.$$

2.5. Another mean of two real numbers a and b is the *root mean square* or *quadratic mean* $\sqrt{(a^2 + b^2)/2}$, common in physics and electrical engineering where it is used to measure magnitude for quantities that may be both positive or negative, such as waveforms. Show that the Cauchy-Schwarz inequality implies the *arithmetic mean-root mean square inequality* for positive numbers a and b:

$$\frac{a+b}{2} \leq \sqrt{\frac{a^2+b^2}{2}}.$$

2.6. Let ABC be a triangle. In Figure 2.3 we saw that if C is a right angle, then the angle bisector of C partitions the square on side AB into two congruent trapezoids. Is the converse true?

CHAPTER 3

Garfield's Trapezoid

> *My mind seems unusually clear and vigorous in Mathematics, and I have considerable hope and faith in the future.*
>
> James A. Garfield

In 1876 a new proof of the Pythagorean theorem appeared in the *New England Journal of Education* (volume 3, page 161). The author of this proof was James Abram Garfield (1831–1881) of Ohio, a member of the United States House of Representatives. The proof was unusual in that it employed a trapezoid constructed from right triangles, an icon we call *Garfield's trapezoid*. In 1880 Garfield was elected the twentieth President of the United States, only to be assassinated four months after his inauguration. He was the last American president to have been born in a log cabin. For more on Garfield's life and his mathematics, see [Hill, 2002].

President James A. Garfield

In this chapter we examine Garfield's proof, and a variety of other results obtained by generalizing the trapezoid and enclosing it in a rectangle.

3.1   The Pythagorean theorem—the Presidential proof

Although his trapezoid resembles the portion of the *Zhou bi suan jing* below a diagonal of the inner square, by all accounts Garfield was unaware of the *Zhou bi suan jing* proof. Garfield's trapezoid leads to an entirely different proof of the Pythagorean theorem, one that is algebraic rather than geometric. Garfield's proof proceeds by computing the area of the trapezoid in Figure 3.1 in two different ways.

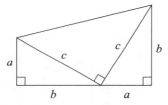

Figure 3.1.

The area of the trapezoid is the product of the base $a + b$ and the average height $(a + b)/2$, i.e., $(a + b)^2/2$. It is also the sum of the areas of the three right triangles, that is, $ab/2 + ab/2 + c^2/2 = (2ab + c^2)/2$. Equating the two expressions, multiplying by two, and simplifying yields

$$(a + b)^2 = 2ab + c^2,$$
$$a^2 + 2ab + b^2 = 2ab + c^2,$$
$$a^2 + b^2 = c^2.$$

3.2   Inequalities and Garfield's trapezoid

The following problem appeared as Problem 3 in the 1969 Canadian Mathematical Olympiad:

Let c be the length of the hypotenuse of a right angle triangle whose other two sides have lengths a and b. Prove that

$$a + b \leq c\sqrt{2}. \tag{3.1}$$

When does equality hold?

The solution follows immediately from the Garfield trapezoid in Figure 3.1. The upper edge of the trapezoid has length $c\sqrt{2}$ (from the Pythagorean theorem just proven), and is at least as long as the base $a + b$. The two are equal if and only if the upper and lower edges are parallel, which occurs if and only if $a = b$.

If we divide both sides of (3.1) by 2 and recall that $c = \sqrt{a^2 + b^2}$ we have

$$\frac{a + b}{2} \leq \sqrt{\frac{a^2 + b^2}{2}} \tag{3.2}$$

with equality if and only if $a = b$. The term on the right side in (3.2) is the root mean square, which we encountered in Challenge 2.5, and (3.2) is the arithmetic mean-root mean square inequality.

3.3 Trigonometric formulas and identities

Since the Garfield trapezoid is constructed from three triangles that share sides, it is remarkably well suited for illustrating a variety of trigonometric formulas and identities. In this section we often adjoin a fourth triangle to the trapezoid, and use the resulting rectangle to explore relationships among the sides and angles of the triangles in the configuration.

We begin by setting $a = 1$ and $b = 2$ in Figure 3.1 and embedding the icon in a 2-by-3 rectangle, as illustrated in Figure 3.2a. This enables us to evaluate certain angles marked $\angle$ in the shaded triangles as arctangents and sums of arctangents.

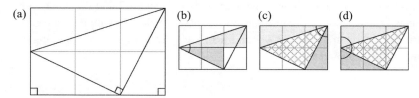

Figure 3.2.

From the marked angles in Figures 3.2b, 3.2c, and 3.2d we obtain

$$\arctan(1/2) + \arctan(1/3) = \pi/4, \tag{3.3}$$

$$\arctan(1) + \arctan(1/2) + \arctan(1/3) = \pi/2, \text{ and}$$

$$\arctan(1) + \arctan(2) + \arctan(3) = \pi \, [\text{Wu, 2003}].$$

The argument can be easily modified to derive *Euler's arctangent identity*: For positive numbers p and q,

$$\arctan(1/p) = \arctan(1/(p+q)) + \arctan(q/(p^2 + pq + 1)). \quad (3.4)$$

See Figure 3.3 [Wu, 2004], and observe that $\alpha = \beta + \gamma$ where $\alpha = \arctan(1/p), \beta = \arctan(1/(p+q))$, and $\gamma = \arctan(q/(p^2 + pq + 1))$.

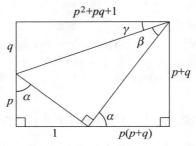

Figure 3.3.

When $p = q = 1$, we obtain (3.3) as a special case of (3.4). See Challenges 3.3 and 3.4 for further arctangent identities.

The usual way to evaluate the trigonometric functions of 15° and 75° is with addition and subtraction formulas, e.g., $\sin(15°) = \sin(45° - 30°)$ or $\sin(60° - 45°)$, $\tan(75°) = \tan(30° + 45°)$, etc. However, they are readily evaluated geometrically using a modified Garfield trapezoid and a procedure adapted from [Hoehn, 2004]. In Figure 3.4a we construct the trapezoid from two isosceles right triangles, one with sides 1, 1, and $\sqrt{2}$ and the other with sides $\sqrt{3}$, $\sqrt{3}$, and $\sqrt{6}$. Since the sides containing the right angle in the gray triangle in Figure 3.4a are $\sqrt{2}$ and $\sqrt{6}$, the gray triangle has acute angles measuring 30° and 60°.

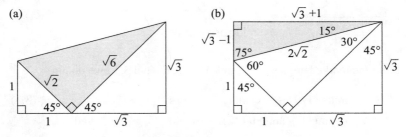

Figure 3.4.

Adjoin a right triangle to the top of the trapezoid to form the rectangle in Figure 3.4b. The acute angles of the adjoined triangle (shaded gray) are 15° and 75°; the sides are $\sqrt{3} - 1$, $\sqrt{3} + 1$, and $2\sqrt{2}$; and hence

$$\sin 15° = \frac{\sqrt{3} - 1}{2\sqrt{2}}, \tan 75° = \frac{\sqrt{3} + 1}{\sqrt{3} - 1}, \text{etc.}$$

The procedure can be modified to illustrate the addition and subtraction formulas for the sine, cosine, and tangent, at least for small positive angles. Let α and β be acute angles whose sum is less than $\pi/2$, and arrange two triangles with acute angle α and one with acute angle β as shown in Figure 3.5.

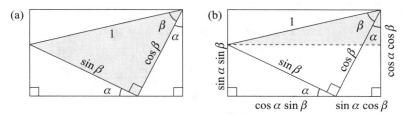

Figure 3.5.

If we let the sides of the gray triangle in Figure 3.5a be $\sin \beta$, $\cos \beta$, and 1, then it is easy to compute the lengths of the sides of the triangles with acute angle α, as shown in Figure 3.5b. Evaluating the lengths of the sides of the gray triangle in Figure 3.5b yields

$$\sin(\alpha + \beta) = \sin \alpha \cos \beta + \cos \alpha \sin \beta,$$
$$\cos(\alpha + \beta) = \cos \alpha \cos \beta - \sin \alpha \sin \beta.$$

The other formulas can be similarly illustrated, see Challenges 3.5 and 3.6.

We conclude this section with two little known identities for sums and products of tangents, again illustrated with modified Garfield trapezoids. In Figure 3.6 we show that for positive angles α, β, and γ such that $\alpha + \beta + \gamma = \pi$ (for instance, the angles in a triangle), we have

$$\tan \alpha + \tan \beta + \tan \gamma = \tan \alpha \tan \beta \tan \gamma.$$

The identity results from equating the two expressions for the vertical height of the rectangle.

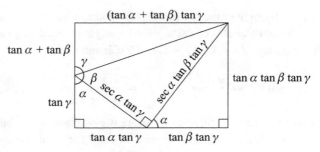

Figure 3.6.

In Figure 3.7 we see a complementary result: for positive angles α, β, and γ such that $\alpha + \beta + \gamma = \pi/2$, we have

$$\tan \alpha \tan \beta + \tan \beta \tan \gamma + \tan \gamma \tan \alpha = 1.$$

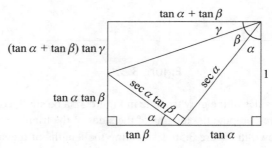

Figure 3.7.

3.4   Challenges

3.1. Use Garfield's trapezoid to show that for any θ, $|\sin \theta + \cos \theta| \leq \sqrt{2}$.

3.2. Use Garfield's trapezoid to illustrate the AM-GM inequality.

3.3. Let $0 < a < b$. Use a diagram similar to Figure 3.2a to show

$$\arctan(a/b) + \arctan((b - a)/(b + a)) = \pi/4.$$

3.4. Let $\{F_n\}_{n=1}^{\infty}$ denote the sequence of Fibonacci numbers, defined by $F_1 = F_2 = 1$ and $F_n = F_{n-1} + F_{n-2}$ for $n \geq 3$. Prove

$$\arctan \frac{1}{F_{2n}} = \arctan \frac{1}{F_{2n+1}} + \arctan \frac{1}{F_{2n+2}}$$

for $n \geq 1$. [Hint: Use Euler's arctangent identity and *Cassini's identity* for the Fibonacci numbers: $F_{k-1}F_{k+1} - F_k^2 = (-1)^k$ for $k \geq 2$. See Challenge 18.2 for a proof of Cassini's identity.]

3.5. Use a diagram similar to Figure 3.4 to illustrate the subtraction formulas for the sine and cosine:

$$\sin(\alpha - \beta) = \sin \alpha \cos \beta - \cos \alpha \sin \beta,$$
$$\cos(\alpha - \beta) = \cos \alpha \cos \beta + \sin \alpha \sin \beta.$$

3.6. Use diagrams similar to Figure 3.4 to illustrate the addition and subtraction formulas for the tangent:

$$\tan(\alpha + \beta) = \frac{\tan \alpha + \tan \beta}{1 - \tan \alpha \tan \beta} \quad \text{and} \quad \tan(\alpha - \beta) = \frac{\tan \alpha - \tan \beta}{1 + \tan \alpha \tan \beta}.$$

3.7. Let a, b, c be the sides of a right triangle, with c the hypotenuse. Show that three right triangles (each similar to the given triangle) with sides a^2, ab, ac; ab, b^2, bc; and ac, bc, c^2; can be arranged to form a Garfield trapezoid that is actually a rectangle, yielding another proof of the Pythagorean theorem.

3.8. An integration technique often presented in calculus is the *Weierstrass substitution*, useful for integrals of rational functions of the sine and cosine. The change of variables $z = \tan(\theta/2)$ yields rational functions of z for $\sin \theta$ and $\cos \theta$. Use the Garfield trapezoid in Figure 3.8 to evaluate $\sin \theta$ and $\cos \theta$ as functions of z.

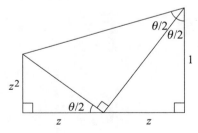

Figure 3.8.

3.9. Given $a, b > 0$ consider for any $k > 0$ the trapezoid in Figure 3.9, where P_k is the midpoint of the slanted edge. Describe the position of P_k as a function of k.

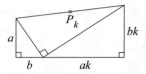

Figure 3.9.

3.10. It is an immediate consequence of the AM-GM inequality (see Section 2.2 or Challenge 3.2) that for x positive, $x + (1/x) \geq 2$. Show that this inequality can be improved to

$$x + \frac{1}{x} \geq \frac{1}{2}\left(\sqrt{x} + \frac{1}{\sqrt{x}}\right)^2 \geq 2. \tag{3.5}$$

CHAPTER **4**

The Semicircle

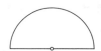

> *The eye is the first circle; the horizon which it forms is the second; and throughout nature this primary figure is repeated without end. It is the highest emblem in the cipher of the world.*
>
> Ralph Waldo Emerson, *Essays*

The semicircle has long played an important role in architecture and art. The Romans employed semicircular arches in their constructions such as the Pont du Gard aqueduct near Nîmes, France (see Figure 4.1a). Similar arches can be found in Romanesque architecture throughout Europe. The Moors also used semicircular arches in their most impressive buildings, such as in the interior of the Mezquita mosque in Córdoba, Spain (see Figure 4.1b). Semicircles also appear in paintings of modern artists such as Wassily Kandinsky (*Semicircle*, 1927) and Robert Mangold (*Semi-Circle I–IV*, 1995).

(a)   (b)

Figure 4.1.

The icon for this chapter is a semicircle, simple and easily constructed with straightedge and compass but nonetheless remarkably useful in

29

geometry and trigonometry. It figures prominently in the work of Greek geometers such as Thales, Euclid, Archimedes, Hippocrates, and Pappus. The
semicircle also appears in the solution of the first optimization problem in
literature, Dido's problem. We also encounter it in visual demonstrations of a
variety of trigonometric identities, and in an exploration of the five Platonic
solids inscribed in a sphere.

4.1   Thales' triangle theorem

Perhaps the first occurrence of the semicircular icon in geometry is in a
theorem named after Thales of Miletus (circa 624–546 BCE), one of the
earliest of the great Greek mathematicians. The theorem also appears as
Proposition 31 in Book III of Euclid's *Elements*. There are two theorems
commonly called "Thales' theorem"—one about a right angle inscribed in a
semicircle, which we call *Thales' triangle theorem*, and another about similar triangles, which we call *Thales' proportionality theorem* and present in
Chapter 5.

Thales' triangle theorem. *A triangle inscribed in a semicircle is a right
triangle.* See Figure 4.2a.

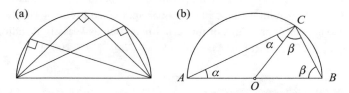

Figure 4.2.

Label the vertices of the triangle as shown in Figure 4.2b, and let O denote
the center of the semicircle. Then $|AO| = |BO| = |CO|$ since each is the
length of a radius, and thus triangles AOC and BOC are isosceles. Hence
$\angle OAC = \angle OCA = \alpha$ and $\angle OBC = \angle OCB = \beta$. The angles of triangle
ABC sum to 180°, so $2\alpha + 2\beta = 180°$, or $\alpha + \beta = 90°$. Thus C is a right
angle as claimed.

Also, $2\alpha + 2\beta = 180°$ implies that $\angle BOC = 2\alpha$.

Thales' triangle theorem is used repeatedly throughout this chapter. The
converse of the theorem also holds, see Challenge 4.1.

4.2   The right triangle altitude theorem and the geometric mean

In Section 2.2 we encountered the *geometric mean* $\sqrt{ab}$ of two positive numbers a and b (and we will meet it again in Sections 13.4 and 18.4). Why is it called geometric? The three numbers a, $\sqrt{ab}$, b are in geometric progression, just as the three numbers a, $(a+b)/2$, b (where $(a+b)/2$ is the arithmetic mean of a and b) are in arithmetic progression, but it is more likely that the answer lies in the purely geometric construction found in Proposition 13 in Book VI of the *Elements* of Euclid, and illustrated in Figure 4.3a.

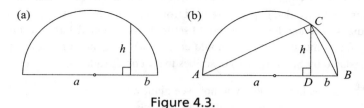

Figure 4.3.

The construction in Figure 4.3a is based on the corollary to Proposition 8 in that Book, which is also known as the

Right triangle altitude theorem: *The length of the altitude to the hypotenuse of a right triangle is the geometric mean of the lengths of the resulting two segments of the hypotenuse.*

Thus in Figure 4.3a we have $h = \sqrt{ab}$.

From Thales' triangle theorem, ABC is a right triangle and the half-chord CD partitions ABC into two triangles ACD and BCD, each similar to ABC. Ratios of corresponding sides in ACD and BCD yields $a/h = h/b$ so that $h = \sqrt{ab}$ is the geometric mean of a and b.

Since $h^2 = ab$, h is the side of a square equal in area to a rectangle with sides a and b, yielding a geometric average of a and b. The arithmetic mean $(a+b)/2$ of a and b appears in Figure 4.3b as the radius of the semicircle. Since the radius to C (not drawn) is at least as long as h, we have another proof of the arithmetic mean-geometric mean inequality $(a+b)/2 \geq \sqrt{ab}$.

When we construct a square equal in area to a given figure, we say we have *squared* the figure. Figure 4.3a shows how to square rectangles. To square a triangle, first construct a rectangle equal in area to the triangle. To square a convex n-gon, first partition it into triangles and square them. By virtue of the Pythagorean theorem, we can construct a square equal in area to two squares, and repeating this procedure yields a square equal in area to any number of squares, which completes the squaring of the n-gon.

4.3   Queen Dido's semicircle

The legend of Dido comes to us from the epic poem *Aeneid*, written by the Roman poet Virgil (70–19 BCE). Dido was a princess from the Phoenician city of Tyre (in present-day Lebanon). Dido fled the city after her brother murdered her husband and arrived in Africa in about 900 BCE, near the bay of Tunis. Dido decided to purchase land from the local leader, King Jarbas of Numidia, so she and her people could settle there. She paid Jarbas a sum of money for as much land as she could enclose with the hide of an ox. Virgil (as translated by Sarah Ruden) describes the scene as follows:

> They came here, where you now see giant walls
> And the rising citadel of newborn Carthage.
> They purchased land, 'as much as one bull's hide
> Could reach around,' and called the place 'the Bull's Hide.'

To obtain as much land as possible, Dido cut the ox skin into thin strips and tied them together, as illustrated in the 17th century woodcut in Figure 4.4. This plot of land would later become the site of the city of Carthage.

Figure 4.4.

This brings us to *Dido's problem*: How should she lay the strips on the ground to enclose as much land as possible? If we assume that the ground is flat and the Mediterranean shore is a straight line, then the optimal solution

is to lay the strips in the shape of a semicircle, which legend tells us is what
Dido did. Our proof is from [Niven, 1981], who attributes it to Jakob Steiner
(1796–1863).

To prove that the semicircle is the optimal solution we first need the elementary result: *Of all triangles with two sides of fixed length and a third side
of arbitrary length, the triangle with the largest area is the right triangle
with the arbitrary side as hypotenuse.* If the fixed sides have lengths a and
b and if θ denotes the angle between them, then the area T of the triangle
equals $(ab \sin \theta)/2$. To maximize T we need only maximize $\sin \theta$, which
occurs if and only if $\theta = 90°$.

Dido's problem is closely related to the *isoperimetric problem* ("isoperimetric" means "having the same perimeter"): Among all closed curves of
a given length, which encloses the greatest area? The difference is that in
Dido's case only the length of the strips of oxhide was fixed, so she was free
to use as much of the straight shoreline as she wished. We call the solution

Dido's theorem. *Suppose a curve C of fixed length and a straight line L
enclose a region. If C is not a semicircle, then it can be replaced by another
curve C' of the same length so that C' and L enclose a region with greater
area. Hence if there is a curve enclosing (with L) a region with maximal
area, it must be a semicircle.*

Let A and B be the points of L which are the endpoints of the curve C, as
illustrated in Figure 4.5a. From Thales' triangle theorem we know that if C
is not a semicircle, then we can find a point P on C where $\angle APB \neq 90°$.
Then the region enclosed by C and L consists of three parts: the region R
enclosed by the line segment AP and C, the region S enclosed by the line
segment BP and C, and the triangle APB denoted by T.

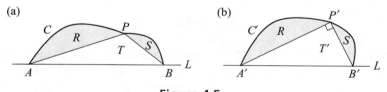

Figure 4.5.

With AP and BP fixed at the point P, slide A and B along L to A' and
B' with $|A'P'| = |AP|$ and $|B'P'| = |BP|$ so that $\angle A'P'B' = 90°$, as
shown in Figure 4.5b. By construction, the shaded areas R and S remain the
same, but the area of T' is greater that the area of T, so that the curve C has

been transformed into a curve C' with the same length as C but enclosing a greater area with L.

As noted in the statement of Dido's theorem, we must assume that a curve enclosing the maximum area exists. The existence of such a curve was proven formally by Karl Theodor Wilhelm Weierstrass (1815–1897).

4.4   The semicircles of Archimedes

The *Book of Lemmas* (or *Liber Asumptorum*) is a collection of fifteen propositions and their proofs attributed to Archimedes (287–212 BCE). It survives today from an Arabic translation by Thabit ibn Qurra (836–901). Several of the propositions in the *Book of Lemmas* concern semicircles and we present four of them, Propositions 3, 4, 8, and 14. Proposition 3 presents a bisection of the vertical line segment in our icon, proposition 8 uses the semicircle to trisect an angle, and propositions 4 and 14 concern area of figures known as an *arbelos* (shoemaker's knife) and a *salinon* (a salt-cellar). See [Heath, 1897] for all the propositions and their proofs.

Archimedes' Proposition 3. *Let AB be the diameter of a semicircle, and let the tangents to it at B and at any other point D on it meet at T. If now DE is drawn perpendicular to AB, and if AT, DE meet at F, then $|DF| = |FE|$.* See Figure 4.6a.

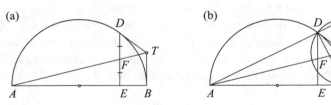

Figure 4.6.

Draw AD, and extend it and BT to meet at H, as shown in Figure 4.6b. Since $\angle ADB$ is a right angle, so is $\angle BDH$. In addition we have $|BT| = |TD|$. Hence T is the center of a semicircle with diameter BH, and $|BT| = |TH|$. Since DE and BH are parallel, it follows that $|DF| = |FE|$.

Archimedes' Proposition 4. *Let P, Q, and R be three points on a line, with Q lying between P and R. Semicircles are drawn on the same side of the line with diameters PQ, QR, and PR. An arbelos is the figure bounded by the three semicircles. Draw the perpendicular to PR at Q, meeting the*

largest semicircle at S. Then the area A of the arbelos equals the area C of
the circle with diameter QS. See Figure 4.7.

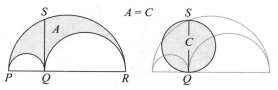

Figure 4.7.

There is a direct algebraic proof of Proposition 4 using the geometric mean, which we present as Challenge 4.2. Here we present a geometric proof [Nelsen, 2002b] based on the Pythagorean theorem. In Proposition 31 of Book VI of the *Elements*, Euclid writes: *In right-angled triangles the figure on the side opposite the right angle equals the sum of the similar and similarly described figures on the sides containing the right angle.* For similar figures on the sides, we now use semicircles. See Figure 4.8.

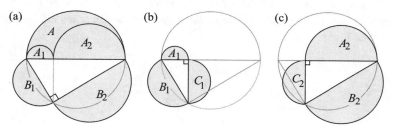

Figure 4.8.

In Figure 4.8a we have $A + A_1 + A_2 = B_1 + B_2$, in Figure 4.8b, $B_1 = A_1 + C_1$, and in Figure 4.8c, $B_2 = A_2 + C_2$. Together the equations yield $A + A_1 + A_2 = A_1 + C_1 + A_2 + C_2$, which simplifies to $A = C_1 + C_2 = C$.

In Proposition 8 Archimedes uses a circle to trisect an angle. We present a slight variation with a semicircle [Aaboe, 1964].

Archimedes' Proposition 8. *In a semicircle with center O and diameter AB, let $\angle AOE$ be the angle to be trisected. Let C be the point on the extension of AB so that the line CE intersects the circle at D with $|CD|$ equal to the radius of the circle. Then $\angle ACE = (1/3)\angle AOE$. See Figure 4.9.*

Let $\alpha = \angle AOE$. Since $|CD| = |OD| = |OE|$, triangles DOE and ODC are isosceles, and hence have equal base angles β and γ, respectively,

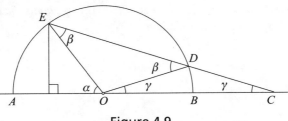

Figure 4.9.

as indicated in Figure 4.9. Hence by the exterior angle theorem (Proposition 32 in Book I of Euclid's *Elements*) $\beta = 2\gamma$ and $\alpha = \beta + \gamma = 3\gamma$.

Figure 4.9 can also be used to derive the triple angle formulas for the sine and cosine. See Section 7.5.

Archimedes' Proposition 14. *Let* P, Q, R, S *be four points on a line (in that order) such that* $PQ = RS$. *Semicircles are drawn above the line with diameters* PQ, RS, *and* PS, *and another semicircle with diameter* QR *is drawn below the line.* A salinon *is the figure bounded by the four semicircles. Let the axis of symmetry of the salinon intersect its boundary at* M *and* N. *Then the area* A *of the salinon equals the area* C *of the circle with diameter* MN. *See Figure 4.10.*

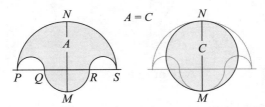

Figure 4.10.

In our proof [Nelsen, 2002a], we use the fact that the area of a semicircle is $\pi/2$ times the area of the inscribed isosceles right triangle (see Figure 4.11).

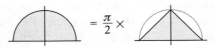

Figure 4.11.

Consequently, the area of the salinon is $\pi/2$ times the area of a square (see Figure 4.12), which equals the area of a circle and proves the proposition.

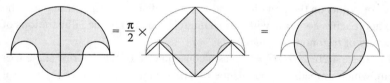

Figure 4.12.

4.5 Pappus and the harmonic mean

The *harmonic mean* of two positive numbers a and b is often defined as the reciprocal of the arithmetic mean of the reciprocals of a and b, i.e., $([(1/a) + (1/b)]/2)^{-1}$, which simplifies to $2ab/(a + b)$. The name harmonic derives from its connection with the harmonic sequence $1, 1/2, 1/3, 1/4, \ldots$, which is important in music. In the harmonic sequence, each term is the harmonic mean of its predecessor and its successor, e.g., $1/3$ is the harmonic mean of $1/2$ and $1/4$.

We have seen how the arithmetic and geometric means of a and b appear as line segments in a semicircle with diameter $a + b$ in Figure 4.3a: the radius is the arithmetic mean $(a + b)/2$ and the half-chord perpendicular to the diameter is the geometric mean $\sqrt{ab}$. Pappus of Alexandria (circa 290–350), in Book III of his *Collection*, added one more line segment to illustrate the harmonic mean as well, as shown in Figure 4.13.

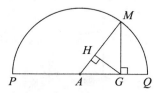

Figure 4.13.

When $|PG| = a$ and $|GQ| = b$, the arithmetic and geometric means are $|AM| = (a+b)/2$ and $|GM| = \sqrt{ab}$ respectively. Since GH is perpendicular to AM, triangles AGM and GHM are similar, so that $|HM|/|GM| = |GM|/|AM|$. Thus the harmonic mean is $|HM| = 2ab/(a + b)$, with $|HM| \leq |GM| \leq |AM|$. Also, the geometric mean of a and b equals the geometric mean of $(a + b)/2$ and $2ab/(a + b)$.

4.6   More trigonometric identities

The trigonometric functions can be expressed both in terms of the coordinates of points on the unit circle and as ratios of sides of right triangles. The two approaches combine with our semicircle icon to produce illustrations of a variety of trigonometric identities.

In Figure 4.2b, $\angle BOC = 2\alpha$ as a consequence of the exterior angle theorem, which immediately yields the two half-angle formulas for the tangent function.

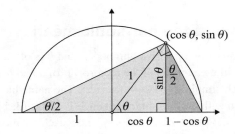

Figure 4.14.

In the light gray triangle in Figure 4.14 we have $\tan(\theta/2) = \sin\theta/(1 + \cos\theta)$, and in the dark gray one we have $\tan(\theta/2) = (1 - \cos\theta)/\sin\theta$ [Walker, 1942]. Figure 4.14 can also be used to establish certain inverse trigonometric identities: see Challenge 4.3.

If we replace θ by 2θ in Figure 4.14, we can obtain the double-angle formulas for the sine and cosine. See Figure 4.15.

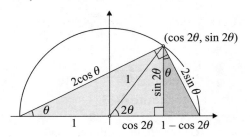

Figure 4.15.

In the light gray triangle, we have $\sin\theta = \sin 2\theta/2\cos\theta$, hence $\sin 2\theta = 2\sin\theta\cos\theta$, $\cos\theta = (1 + \cos 2\theta)/2\cos\theta$ or $\cos 2\theta = 2\cos^2\theta - 1$. In the dark gray triangle, we have $\sin\theta = (1 - \cos 2\theta)/2\sin\theta$, so that $\cos 2\theta =$

$1 - 2 \sin^2 \theta$ [Woods, 1936]. See Challenge 4.4 for another derivation of the formulas.

In Challenge 3.8 we encountered the Weierstrass substitution, useful for integrals of rational functions of the sine and cosine. The change of variables $z = \tan(\theta/2)$ yields rational functions of z for $\sin \theta$ and $\cos \theta$, as illustrated in Figure 4.16.

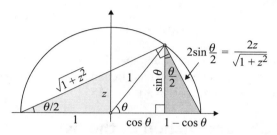

Figure 4.16.

Since $z = \tan(\theta/2)$, the sides of the light gray triangle are 1, z, and $\sqrt{1 + z^2}$, and consequently the hypotenuse of the dark gray triangle is $2 \sin(\theta/2) = 2z/\sqrt{1 + z^2}$. Since the shaded triangles are similar, we have

$$\frac{\sin \theta}{2z/\sqrt{1 + z^2}} = \frac{1}{\sqrt{1 + z^2}} \text{ and } \frac{1 - \cos \theta}{2z/\sqrt{1 + z^2}} = \frac{z}{\sqrt{1 + z^2}},$$

and hence $\sin \theta = 2z/(1 + z^2)$ and $\cos \theta = (1 - z^2)/(1 + z^2)$ [Deiermann, 1998].

4.7 Areas and perimeters of regular polygons

There is a nice relationship between the perimeter of a regular polygon with n sides (an n-gon) and the area of a regular polygon with $2n$ sides when both are inscribed in a circle of radius r. If we let s_n denote the side length, $P_n = ns_n$ the perimeter, and A_n the area of a regular n-gon, then the area $A_{2n}/2n$ of the shaded triangle in Figure 4.17 is equal to one-half its base r times its altitude $s_n/2$, i.e., $A_{2n}/2n = rs_n/4$.

Hence the area of the $2n$-gon is $A_{2n} = rns_n/2 = rP_n/2$, i.e., half the product of the circumradius and perimeter of the n-gon. For example, the perimeter of a regular hexagon inscribed in a circle of radius r is $P_6 = 6r$, and hence the area of a regular dodecagon inscribed in the same circle is $A_{12} = rP_6/2 = 3r^2$.

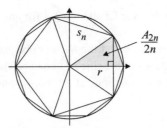

Figure 4.17.

As a consequence we can conclude that the area of a circle is equal to one-half the product of its radius and its circumference, a fact known to the Indian mathematician Bhāskara (circa 1114–1185).

4.8   Euclid's construction of the five Platonic solids

The last of the thirteen Books of the *Elements* of Euclid concerns the five Platonic solids—the tetrahedron, cube, octahedron, dodecahedron, and icosahedron. After numerous Propositions about the individual solids Proposition 18, the final Proposition of the final Book, provides a construction for the edge length of each of the five solids when inscribed in the same sphere. The construction is remarkably simple, based on a semicircle and several segments perpendicular to its diameter. Figure 4.18 is a simplified version of a diagram that appears in many translations of the *Elements*.

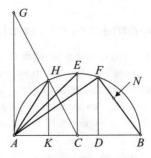

Figure 4.18.

Draw a semicircle with AB as diameter, and locate points C and D on AB so that $|AC| = |CB|$ and $|AD| = 2|DB|$. Draw AG perpendicular to

AB with $|AG| = |AB|$, let GC intersect the semicircle at H, and draw HK perpendicular to AB. Draw CE and DF perpendicular to AB, draw AH, AE, AF, and BF, and locate N on BF so that $|BF| = \phi|BN|$ where $\phi = (1 + \sqrt{5})/2$ is the golden ratio, the positive root of the quadratic equation $\phi^2 = \phi + 1$. (Euclid describes how to do this in Proposition II.11.) Then

> AF is the edge of the tetrahedron,
> BF is the edge of the cube,
> AE is the edge of the octahedron,
> BN is the edge of the dodecahedron, and
> AH is the edge of the icosahedron.

We justify the last claim and leave the others to Challenge 4.8. The twelve vertices of an icosahedron with edge s are determined by three s-by-$s\phi$ *golden rectangles*. See Figure 4.19.

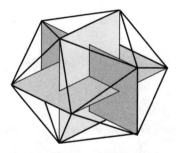

Figure 4.19.

If $|AB| = d$ in Figure 4.18, then triangle geometry yields $|AH| = d/\sqrt{2 + \phi}$. The diagonal of the golden rectangle is also the diameter d of the sphere, and hence $d = \sqrt{s^2 + s^2\phi^2} = s\sqrt{2 + \phi}$. It now follows that $s = d/\sqrt{2 + \phi} = |AH|$, as required.

4.9   Challenges

4.1. Prove the converse of Thales' triangle theorem: the hypotenuse of a right triangle is a diameter of its circumcircle.

4.2. Prove Proposition 4 from the *Book of Lemmas* using the fact that, in Figure 4.7, $|QS|$ is the geometric mean of $|PQ|$ and $|QR|$.

4.3. Many identities for inverse trigonometric functions are equivalent to identities for trigonometric functions. Establish the following identities by relabeling sides and angles in Figure 4.14:

(a) $\dfrac{\arcsin x}{2} = \arctan\dfrac{x}{1 + \sqrt{1 - x^2}} = \arctan\dfrac{1 - \sqrt{1 - x^2}}{x}$;

(b) $\dfrac{\arccos x}{2} = \arctan\dfrac{1 - x}{\sqrt{1 - x^2}} = \arctan\dfrac{\sqrt{1 - x^2}}{1 + x} = \arctan\sqrt{\dfrac{1 - x}{1 + x}}$;

(c) $\dfrac{\arctan x}{2} = \arctan\dfrac{x}{1 + \sqrt{1 + x^2}} = \arctan\dfrac{\sqrt{1 + x^2} - 1}{x}$.

4.4. Use Figure 4.20 to derive the double angle formulas for the sine and cosine. (Hints: Compute the area of the shaded triangle in two ways, and express the length of the chord in two ways.)

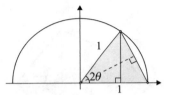

Figure 4.20.

4.5. Using the semicircle icon show that $\arctan x \le \arcsin x$ for x in $[0,1]$.

4.6. Show that the area of the shaded region in Figure 4.21 depends only on a.

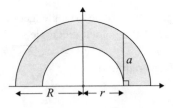

Figure 4.21.

4.7. If we draw semicircles on the legs of a right triangle inscribed in a semicircle, we create a *lune* on each leg, the region inside the small semicircle and outside the large one, as illustrated in light gray in Figure 4.22. Show that the combined area of the two lunes equals the area of the right triangle, a fact known to Hippocrates of Chios (circa 470–410 BCE).

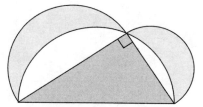

Figure 4.22.

4.8. In Figure 4.18, show that AF, BF, AE, and BN are, respectively, the edges of the tetrahedron, cube, octahedron, and dodecahedron inscribed in a sphere of diameter AB.

4.9. Another curve bounded by three semicircular arcs, like the arbelos, is the heart-shaped curve called the *cardioid of Boscovich* (Roger Boscovich, 1711–1787) illustrated in Figure 4.23. Show that any line through the cusp bisects its perimeter.

Figure 4.23.

CHAPTER **5**

Similar Figures

In the physical world, one cannot increase the size or quantity of anything without changing its quality. Similar figures exist only in pure geometry.

Paul Valéry

Similar figures not only exist in geometry, they are pervasive. Our icon for this chapter on similar figures is a pair of similar triangles. Equality of ratios of corresponding sides of similar triangles—Thales' proportionality theorem—is the key to exploring similarity of figures in geometry. Consequences of the theorem include indirect measurement, trigonometric identities, geometric sequences and series, etc.

But what are "similar" figures? Intuitively, similar objects are objects with the same shape, or which are congruent after suitable scaling. We will need to be more specific when dealing with mathematical (i.e., geometric) objects such as triangles and other polygons. Similar objects abound in everyday life—both man-made, such as the Russian matryoshka dolls, and in nature, such as the mother elephant and her baby—as seen in Figure 5.1. Similarities also appear in charts, maps, photographs, flowers, leaves, toy cars and trains, dollhouses, etc.

Figure 5.1.

We begin by proving Thales' theorem and exploring some of its consequences for right and general triangles. We use similar triangles to illustrate
trigonometric identities, sum a geometric series, prove Menelaus's theorem,
and conclude with applications to replicating tiles.

Similarity in literature

In 1726 Jonathan Swift (1667–1745) published his classic novel, *Gulliver's Travels*. In the first part the protagonist Lemuel Gulliver is ship-
wrecked on the island of Lilliput. Shortly thereafter Gulliver encounters
"a human Creature not six Inches high, with a Bow and Arrow in his
Hands." The inhabitants of Lilliput were in all respects similar to Englishmen, only one-twelfth the size. In the second part Gulliver lands in
the country of Brobdingnag, where the inhabitants are again similar to
Englishmen, only about twelve times as large. Swift used the size differ-
ences between Gulliver and the inhabitants of Lilliput and Brobgingnag
to make social criticism and political commentary.

Charles Lutwidge Dodgson (1832–1898), better known as Lewis
Carroll, published *Alice's Adventures in Wonderland* in 1865. In the first
chapter, Alice, who was about ten years old, finds a little bottle with
the label "DRINK ME," and does so. She shrinks to only ten inches in
height, but in all other aspects, she is still Alice. Alice wonders how long
she will continue to shrink. "It might end, you know, in my going out
altogether, like a candle," a quote some mathematicians say is Carroll's
reference to the concept of a limit.

5.1   Thales' proportionality theorem

Similar triangles are triangles with the same shape, but not necessarily the
same size. To be more precise, the corresponding angles of similar triangles
are equal, so that they are congruent subject to suitable scaling. *Thales' proportionality theorem* states that corresponding sides of similar triangles are
proportional. We first prove the theorem for the legs of similar right triangles.

Given a pair of similar right triangles (a, b, c) and (a', b', c') such as those
in Figure 5.2a, we can construct the rectangle in Figure 5.2b. Its diagonal
divides it into two congruent right triangles with equal areas. Since the areas
of the similarly shaded triangles are the same, the areas ab' and $a'b$ of the
white rectangles must also be equal, from which it follows that $a'/a = b'/b$.

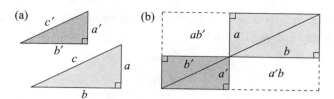

Figure 5.2.

To extend the theorem to include c and c' we arrange the triangles as shown in Figure 5.3a and draw the shaded parallelogram in Figure 5.3b. Computing the area of the parallelogram in two ways yields $a'c = ac'$, equivalent to $a'/a = c'/c$, and hence $a'/a = b'/b = c'/c$.

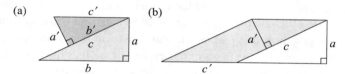

Figure 5.3.

Perhaps one of the simplest proofs of the Pythagorean theorem uses only the equal ratio of sides of similar right triangles. See Figure 5.4, where we have used the altitude h to the hypotenuse c to partition the right triangle into two triangles both similar to the original. Let x and y be the indicated segments of the hypotenuse of triangle (a, b, c).

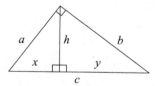

Figure 5.4.

Triangles (a, b, c), (x, h, a) and (h, y, b) are similar, so $x/a = a/c$ and $y/b = b/c$ and thus $x = a^2/c$ and $y = b^2/c$. Since $x + y = c$ it follows that $a^2 + b^2 = c^2$. Also, $x/y = (a/b)^2$ and $xy = (ab/c)^2$.

Similar right triangles can also be used to establish

The Reciprocal Pythagorean theorem. *If a and b are the legs of a right triangle and h the altitude to the hypotenuse c, then $1/a$ and $1/b$ are the legs of a similar right triangle with hypotenuse $1/h$ and hence $(1/a)^2 + (1/b)^2 = (1/h)^2$.*

See Figure 5.5 for a proof that only requires expressing the area of the original triangle as $ab/2 = ch/2$.

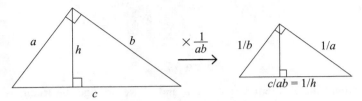

Figure 5.5.

Similar right triangles can be used to illustrate the formula for the sum of a geometric series with positive first term a and common ratio $r < 1$ [Bivens and Klein, 1988]. Draw the large white right triangle in Figure 5.6 similar to the small gray one, and partition it into congruent trapezoids using the parallel gray segments. The bases of the trapezoids are thus the terms of the series, and since the small gray right triangle is similar to the large one, we have

$$\frac{a + ar + ar^2 + \cdots}{1} = \frac{a}{1 - r}.$$

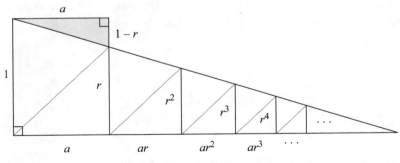

Figure 5.6.

We now extend Thales' proportionality theorem to arbitrary pairs of similar triangles. Suppose ABC and $A'B'C'$ are similar triangles with angles $A = A'$, $B = B'$, $C = C'$, and with the sides labeled a, b, c, a', b', c' in the customary manner. Without loss of generality assume that C is the largest angle of ABC (or one of the largest if there is more than one), and draw the altitude CD from C to AB, and the altitude $C'D'$ from C' to $A'B'$.

Since A and B are acute, D lies between A and B, and similarly D' lies between A' and B'. See Figure 5.7. Let $h = |CD|$, $h' = |C'D'|$, $c_1 = |AD|$, $c'_1 = |A'D'|$, $c_2 = |BD|$, and $c'_2 = |B'D'|$.

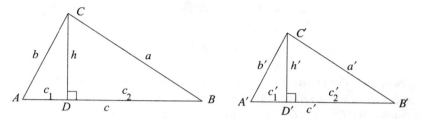

Figure 5.7.

Since triangles ACD and BCD are similar, respectively, to triangles $A'C'D'$ and $B'C'D'$, we have

$$\frac{c'_1}{c_1} = \frac{b'}{b} = \frac{h'}{h} = \frac{a'}{a} = \frac{c'_2}{c_2}.$$

It now follows that $c'_1/c'_2 = c_1/c_2$, so that when 1 is added to both sides we have $c'/c'_2 = c/c_2$, or equivalently, $c'/c = c'_2/c_2$, Hence $a'/a = b'/b = c'/c$ as desired.

In Chapter 1 we used the bride's chair—three squares on the sides of right triangle—to illustrate and prove the Pythagorean theorem. As a consequence of Proposition 31 in Book VI of Euclid's *Elements*, any set of three similar figures can be used. For example, we have semicircles in Section 4.4 and equilateral triangles in Challenge 8.7.

The following identity for positive numbers a, b, c, d will be useful in the next application in this section: If $a/b = c/d \neq 1$, then

$$\frac{a+b}{a-b} = \frac{c+d}{c-d}.$$

First assume without loss of generality that $a > b$ and $c > d$ and consider similar right triangles with legs a, c and b, d as shown in Figure 5.8a.

Using two copies of the smaller triangle and one copy of the larger construct the triangle in Figure 5.8b. Thus the small dark gray right triangle with legs $a - b$ and $c - d$ is similar to the large triangle with legs $a + b$ and $c + d$, so we have the desired result.

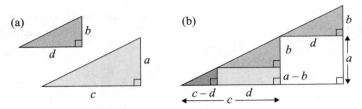

Figure 5.8.

Now consider the trapezoid $ABCD$ in Figure 5.9. Let E denote the intersection of the diagonals AC and BD, and let S and T be the areas of triangles BCE and ADE, respectively. If K denotes the area of $ABCD$, then $K = (\sqrt{T} + \sqrt{S})^2$.

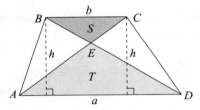

Figure 5.9.

To prove this somewhat surprising result, we first note that since BCE and ADE are similar, $T/S = (a/b)^2$, or $a/b = \sqrt{T}/\sqrt{S}$. Hence, by our previous result, $(a+b)/(a-b) = (\sqrt{T}+\sqrt{S})/(\sqrt{T}-\sqrt{S})$. Computing the area of triangle ABE (or CDE) in two ways yields $ah/2 - T = bh/2 - S$ and hence $(a - b)h/2 = T - S$, so that

$$K = \frac{a+b}{2}h = \frac{a-b}{2}h \cdot \frac{a+b}{a-b} = (T-S)\frac{\sqrt{T}+\sqrt{S}}{\sqrt{T}-\sqrt{S}} = (\sqrt{T}+\sqrt{S})^2,$$

as claimed.

Reduction compasses

A *reduction compass* is a drafting instrument, developed in Italy in the 16th century, to reproduce drawings at a smaller or larger scale. It consists of two intersecting legs with a movable center. The distances between opposite pairs of points thus form simple ratios such as 1:3 or 1:5.

In Figure 5.10 we see an example in brass from the 17th century in the Institute and Museum of the History of Science in Florence.

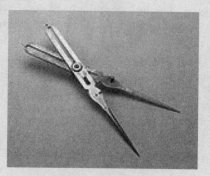

Figure 5.10.

We conclude this section with another proposition from Archimedes' *Book of Lemmas* [Heath. 1897], which we first encountered in Chapter 4. The proof uses properties of similar triangles.

Archimedes' Proposition 1. *If two circles touch at A, and if BD and EF are parallel diameters in them, ADF is a straight line.* See Figure 5.11.

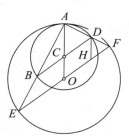

Figure 5.11.

Let O and C be the centers of the circles, and extend OC to meet the circles at A. Draw DH parallel to AO meeting OF at H. Since $|OH| = |CD| = |CA|$ and $|OF| = |OA|$, we have $|HF| = |CO| = |DH|$. So triangles ACD and DHF are similar isosceles triangles, and hence $\angle ADC = \angle DFH$. Adding $\angle CDF$ to both yields

$$\angle ADC + \angle CDF = \angle CDF + \angle DFH = 180°$$

and hence ADF is a straight line.

The Pantograph

A *pantograph* is a mechanical linkage for reproducing drawings at a
larger or smaller scale. The linkage has a fixed point and two movable
points, one with a pointer to trace a drawing and the other with a writing implement such as a pen or pencil to produce a similar image. It
was invented by Christoph Scheiner (1573–1650) in about 1603. The
illustration in Figure 5.12 is from Scheiner's 1631 publication *Panto-
graphice*.

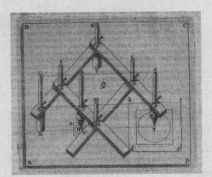

Figure 5.12.

5.2   Menelaus's theorem

The following theorem, commonly attributed to Menelaus of Alexandria
(circa 70–140 CE), provides a criterion for establishing the collinearity of
three points in the plane. There are many proofs, ours uses Thales' proportionality theorem.

Menelaus's theorem. *If X, Y, Z are points on the sides BC, CA, and AB of
triangle ABC (suitably extended) respectively, are collinear, then*

$$\frac{|BX|}{|CX|} \cdot \frac{|CY|}{|AY|} \cdot \frac{|AZ|}{|BZ|} = 1. \tag{5.1}$$

*Conversely, if (5.1) holds for points X, Y, Z on the three sides (extended if
necessary), then the points are collinear.*

  Assume X, Y, Z are collinear. Let h, j, and k are the perpendicular distances from A, B, C, respectively to the line determined by X, Y, and Z, as
shown in Figure 5.13. Using similarity of right triangles we have

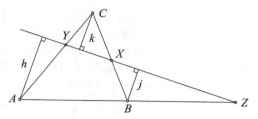

Figure 5.13.

$$\frac{|BX|}{|CX|} = \frac{j}{k}, \frac{|CY|}{|AY|} = \frac{k}{h}, \text{ and } \frac{|AZ|}{|BZ|} = \frac{h}{j},$$

and (5.1) follows. Exactly one or all three of the sides must be extended in order for X, Y, Z to be collinear.

Conversely, assume (5.1) holds. Choose two of the points, say X and Y, and consider the point Z' of intersection of XY and AB. Then we have $|AZ'|/|BZ'| = |AZ|/|BZ|$. Subtracting 1 from both sides yields $|AZ'|/|AB| = |AZ|/|AB|$ so that $|AZ'| = |AZ|$ and $Z' = Z$.

In the next chapter we present Ceva's theorem, a result that complements Menelaus' theorem. While Menelaus's theorem provides a criterion for the collinearity of points on the sides of a triangle, Ceva's theorem gives a similar criterion for the concurrency of lines through the vertices of the triangle.

5.3   Reptiles

Reptiles (for *replicating tiles*) are shapes called *tiles*, usually polygons, that can be joined together to form a larger version of the tile. If n copies of the reptile can form a larger version, it is called *rep-n*. For example, every triangle is a rep-4 and rep-9 reptile, as seen in Figure 5.14, and it is easy to show that every triangle is rep-k^2 for every positive integer k.

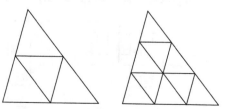

Figure 5.14.

Certain triangles are rep-n reptiles for other values of n. The isosceles right triangle is rep-2, the 30°-60°-90° right triangle is rep-3, and the right triangle with one leg twice the other is rep-5, as illustrated in Figure 5.15. It is known [Snover et al., 1991] that if a triangle is a rep-n reptile, then n is a square, a sum of two squares, or three times a square.

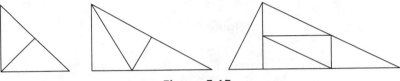

Figure 5.15.

The only regular polygons that are reptiles are the equilateral triangle and the square. However, many polygons are reptiles. For example, every parallelogram is a reptile. An interesting class of polygons are the *polyominoes*, a generalization of dominoes. A polyomino is a union of unit squares in which each square shares at least one edge with another square. See Figure 5.16 for illustrations of a domino, two types of *trominoes* (straight and L), and five types of *tetrominoes* (straight, square, T, skew, and L). Polyominoes, like dominoes, can be rotated and turned over, and are still considered to be of the same type.

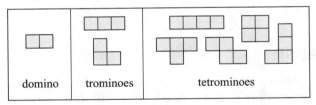

Figure 5.16.

All straight and square polyominoes are reptiles. The L-tromino and the T- and L-tetrominos are reptiles, as seen in Figure 5.17.

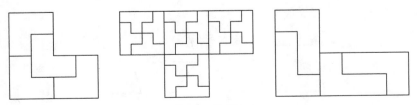

Figure 5.17.

The illustration of the T-tetromino as a rep-16 reptile suggests the following criterion for a polyomino to be a reptile: if the polyomino tiles a square, then it is a reptile. Indeed, if the polyomino tiles a rectangle, it is a reptile. To see this note that mn copies of an $m \times n$ rectangle form a square mn units on a side, and arrange the squares to form a larger version of the polyomino.

Polyiamonds are constructed from equilateral triangles analogous to the way polyominos are constructed from squares. In Figure 5.18 we see the moniamond, diamond, triamond, three types of tetriamonds, and four types of pentiamonds.

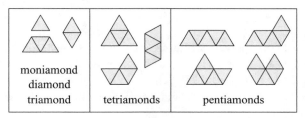

Figure 5.18.

Many of the polyamonds are reptiles. Both the triamond and the *sphinx hexiamond* are rep-k^2 reptiles, illustrated in Figure 5.19 for $k = 2$ and 3 (there are twelve types of hexiamonds and at least three others are reptiles). For more on polyominoes and polyiamonds, see [Martin, 1991].

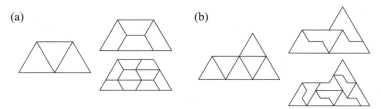

Figure 5.19.

Perfect tilings

In a tiling by reptiles, all the small figures are congruent and similar to the large figure. In a *perfect tiling*, all the small figures are similar to the large one but no two are congruent. Every non-isosceles right triangle admits a perfect tiling with two tiles, as seen in Figure 5.20a. Many other triangles admit perfect tilings, such as the one in Figure 5.20b with six tiles.

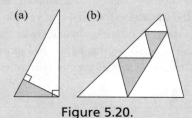

Figure 5.20.

The only known polygons with six or fewer sides which admit perfect tilings with just two tiles are the non-isosceles right triangles above, and the *golden bee* in Figure 5.21 [Scherer, 2010]. It received this name because $r = \sqrt{\phi}$ where ϕ is the golden ratio, and the tile is in the shape of the letter "b".

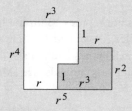

Figure 5.21.

5.4 Homothetic functions

Two figures in the plane are *homothetic* if one is an expansion or contraction of the other with respect to a fixed point in the plane. Thus any pair of homothetic figures must be similar. We encountered homothetic triangles in Section 1.4 when we extended the Vecten configuration. We now consider functions whose graphs are homothetic with respect to the origin.

Let f and g be two functions whose common domain is the set of reals. Then f and g are *homothetic with respect to the origin* (or simply *homothetic*) if there exists a positive constant k, $k \neq 1$, such that for any point

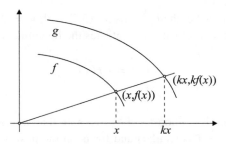

Figure 5.22.

$(x, f(x))$ on the graph of f, the point $(kx, kf(x))$ lies on the graph of g. See Figure 5.22. Consequently we have

$$g(kx) = kf(x), \tag{5.2}$$

or equivalently, replacing x by x/k, the explicit formula

$$g(x) = kf(x/k). \tag{5.3}$$

For example, two rectangular hyperbolas $f(x) = a/x$ and $g(x) = b/x$, $ab > 0$, are homothetic with $k = \sqrt{b/a}$ and two parabolas $f(x) = ax^2$ and $g(x) = bx^2$, $ab > 0$, are homothetic with $k = a/b$.

For another example, let $f(x) = \sin x$ and $g(x) = \sin x \cos x$. Then f and g are homothetic with $k = 1/2$ since

$$g(x) = \sin x \cos x = \frac{1}{2}(2 \sin x \cos x) = \frac{1}{2} \sin 2x = \frac{1}{2} f(2x) = \frac{1}{2} f\left(\frac{x}{1/2}\right).$$

See Figure 5.23, where the gray dashed lines illustrate $2g(x) = f(2x)$ for a couple of values of x.

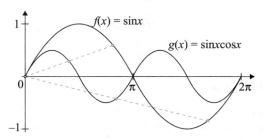

Figure 5.23.

It is obvious that the linear function $f(x) = ax$ is homothetic with itself, what we will call *self-homothetic*. Are there other continuous functions that are self-homothetic?

If f is self-homothetic, then f satisfies (5.2) with $g = f$, i.e., for some positive constant $k \neq 1$ and all x, f satisfies the functional equation

$$f(kx) = kf(x) \qquad\qquad (5.4)$$

with $f(0) = 0$.

To find the general solution to (5.4), we note that we can assume $0 < k < 1$ (the case $k > 1$ is similar) and focus on the positive real line. The positive and negative integer powers of k partition $(0, \infty)$, so we can define f as follows. Let h be an arbitrary continuous function on $[k, 1]$ such that $h(k) = kh(1)$, and let $f = h$ on $[k, 1]$. For x in $[1, 1/k]$ we have $1 \leq x \leq 1/k$, i.e., $k \leq kx \leq 1$, so $h(kx) = f(kx) = kf(x)$, and so on this interval we set $f(x) = (1/k)h(kx)$. Similarly, in $[1/k, 1/k^2]$ we set $f(x) = (1/k^2)h(k^2x)$, and so on. In $[k^2, k]$ we have $k^2 \leq x \leq k$, i.e., $k \leq x/k \leq 1$, so $h(x/k) = f(x/k) = (1/k)f(x)$, and so on this interval we set $f(x) = kh(x/k)$, etc. The condition $h(k) = kh(1)$ insures that f is continuous on $(0, \infty)$.

For example, let $k = 1/2$ and $h(x) = |x - 2/3|$ on $[1/2, 1]$. The graph of f on the interval $(0,2]$ is given in Figure 5.24. The portion of the graph with the heavier line segments is the graph of h.

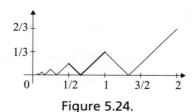

Figure 5.24.

5.5   Challenges

5.1. Let (a, b, c) be a right triangle with hypotenuse c, and let (a', b', c') be a right triangle similar to (a, b, c). Prove $aa' + bb' = cc'$.

5.2. A point P is chosen inside triangle ABC, and lines are drawn through P parallel to the sides of the triangle, as illustrated in Figure 5.25. Let

$a = |BC|, b = |AC|, c = |AB|$, and let a', b', c', be the lengths of the middle segments of each side. Show that

$$\frac{a'}{a} + \frac{b'}{b} + \frac{c'}{c} = 1.$$

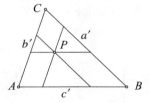

Figure 5.25.

5.3. Given two points P and Q on a circle, consider the chord PQ, the tangent line t at P, and the line QR perpendicular to t, as shown in Figure 5.26. Show that $|PQ|$ is the geometric mean of $|QR|$ and the diameter of the circle.

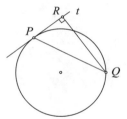

Figure 5.26.

5.4. The diagonals of a square $ABCD$ meet at E, and the bisector of $\angle CAD$ crosses DE at G and meets CD at F, as seen in Figure 5.27. Prove that $|FC| = 2|GE|$.

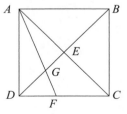

Figure 5.27.

5.5. Let ABC be a triangle inscribed in a circle and consider a point D on the arc BC. Let E be the intersection of BC with AD. When are triangles ABD and BDE similar?

5.6. There are 12 pentominoes, of which the four shown in Figure 5.28— the I, L, P and Y pentominoes—are reptiles. Show that the I and P pentominoes are rep-4, while the L and Y pentominoes are rep-100 pentominoes.

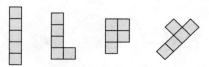

Figure 5.28.

5.7. Can two exponential functions $f(x) = e^{ax}$ and $g(x) = e^{bx}$ with $a \neq b$ be homothetic?

CHAPTER 6

Cevians

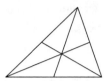

*Arithmetic! Algebra! Geometry! Grandiose trinity! Lumi-
nous triangle! Whoever has not known you is without sense!*
Comte de Lautréamont (1846–1870)

A *cevian* is a line that connects a vertex of a triangle to a point on the opposite side (extended if necessary) of the triangle. Familiar cevians are the
medians, angle-bisectors, and altitudes, but there are many others. The name
comes from the Italian mathematician Giovanni Ceva (1647–1734), and in
the next section we prove Ceva's theorem, providing a necessary and sufficient condition for three cevians to intersect at one point.

Our icon shows a triangle and three concurrent cevians. The point of intersection of the cevians yields a *center* of the triangle. Examples of centers for
the cevians mentioned above are the *centroid* (for the medians), the *incenter*
(for the angle-bisectors), and the *orthocenter* (for the altitudes). In Chapter 1
we encountered two triangle centers, the *Vecten point* and the *Lemoine point*
of the triangle. The corresponding cevians are shown in Figures 1.11 and
1.13a. In Chapter 8 we will see the *Fermat point*. The *Encyclopedia of Triangle Centers* [Kimberling, 2010] is an on-line listing of centers and their properties, and currently (as of 2010) has over 3500 different triangle centers.

Triangles with cevians appear in some everyday objects, commercial logos, highway signs, and yachting club flags, as illustrated in Figure 6.1.

Figure 6.1.

In this chapter we use Ceva's theorem and a closely related result, Stewart's theorem, to examine some properties of the best-known cevians—medians, angle-bisectors, and altitudes. We also examine properties of triangles formed by concurrent cevians. Since not all cevians are concurrent, we conclude with some results about non-concurrent cevians.

Cevians and roof trusses

A *roof truss* is a stable and strong wooden or metal frame to support the roof of a building. Adding cevians enhances the strength and stability by creating smaller triangles within the truss. There are many truss designs. The six ancient ones in Figure 6.2 have different structural properties to meet the demands of the width of the span and the weight of the roof.

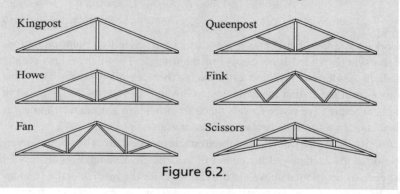

Figure 6.2.

6.1 The theorems of Ceva and Stewart

Ceva's theorem was known long before Giovanni Ceva published a proof in 1678. It was proved by Yusuf al-Mu'taman ibn Hud in the 11th century, and may have been known to the ancient Greeks. Although the theorem is not well known today, it is easy to prove and useful for establishing better-known results.

Ceva's theorem. *Let ABC be a triangle, and X, Y, Z points on sides BC, CA, AB as shown in Figure 6.3. Let X partition BC into segments of length a_b and a_c, and similarly for the other two sides as shown in the figure. The cevians AX, BY, CZ are concurrent if and only if*

$$\frac{a_b}{a_c} \cdot \frac{b_c}{b_a} \cdot \frac{c_a}{c_b} = 1. \tag{6.1}$$

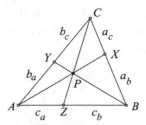

Figure 6.3.

In our proof we need the following simple result: if x, y, z, and t are four real numbers such that $x > z > 0$, $y > t > 0$, and $x/y = z/t$, then

$$\frac{x}{y} = \frac{z}{t} = \frac{x-z}{y-t}. \tag{6.2}$$

The proof of (6.2) follows immediately from Figure 6.4.

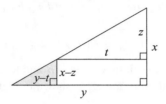

Figure 6.4.

Let $[DEF]$ denote the area of a triangle DEF, and assume that the three cevians AX, BY, CZ in Figure 6.3 are concurrent at P. Since $\triangle ACZ$ and $\triangle BCZ$ have a common altitude, as do $\triangle APZ$ and $\triangle BPZ$, ratios of areas are the same as ratios of bases so that

$$\frac{c_a}{c_b} = \frac{[ACZ]}{[BCZ]} = \frac{[APZ]}{[BPZ]}.$$

Since $[ACZ] > [APZ]$ and $[BCZ] > [BPZ]$, (6.2) yields

$$\frac{c_a}{c_b} = \frac{[ACZ] - [APZ]}{[BCZ] - [BPZ]} = \frac{[ACP]}{[BCP]}. \tag{6.3}$$

Analogously we have

$$\frac{a_b}{a_c} = \frac{[ABP]}{[ACP]} \quad \text{and} \quad \frac{b_c}{b_a} = \frac{[BCP]}{[ABP]}. \tag{6.4}$$

Multiplying the equations in (6.3) and (6.4) yields (6.1).

Conversely, assume that (6.1) holds, and let P be the intersection of the cevians AX and BY. Draw the cevian CZ' through P. Since AX, BY, and CZ' are concurrent, we have

$$\frac{a_b}{a_c} \cdot \frac{b_c}{b_a} \cdot \frac{|AZ'|}{|BZ'|} = 1$$

and since (6.1) holds we have $c_b/c_a = |BZ'|/|AZ'|$. Adding 1 to both sides yields $c_a = |AZ'|$ and hence Z' coincides with Z so that the three cevians AX, BY, and CZ are concurrent.

The following theorem expresses the length of a cevian in terms of the sides of the triangle and the segments on the sides. It is named for Matthew Stewart (1717–1785), a Scottish mathematician who published the result in 1746. The first proof was given by another Scot, Robert Simson (1687–1768), in 1751.

Stewart's theorem. *Let ABC be a triangle with sides a, b, and c. If the cevian CZ divides AB into two segments of length c_a and c_b, then $|CZ|$ satisfies*

$$a^2 c_a + b^2 c_b = c \left(|CZ|^2 + c_a c_b \right). \tag{6.5}$$

To find $|CZ|$ we draw the altitude h_c to AB and let z denote the distance between the foot of the altitude and Z, as shown in Figure 6.5.

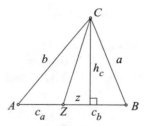

Figure 6.5.

From the Pythagorean theorem, we have

$$a^2 = h_c^2 + (c_b - z)^2 = h_c^2 + c_b^2 + z^2 - 2c_b z = |CZ|^2 + c_b^2 - 2c_b z, \tag{6.6}$$
$$b^2 = h_c^2 + (c_a + z)^2 = h_c^2 + c_a^2 + z^2 + 2c_a z = |CZ|^2 + c_a^2 + 2c_a z. \tag{6.7}$$

Multiplying (6.6) by c_a, (6.7) by c_b, then adding the results and, recalling that $c = c_a + c_b$, yields (6.5). Analogous results hold for $|AX|$ and $|BY|$ in Figure 6.3.

Armed with the theorems of Ceva and Stewart, we now consider some of the better known cevians.

6.2 Medians and the centroid

The medians of a triangle, whose lengths are commonly denoted m_a, m_b, and m_c, are the cevians drawn to the midpoints of the sides. Thus $a_b = a_c = a/2$, $b_a = b_c = b/2$, $c_a = c_b = c/2$ in Figure 6.3, and hence the medians are concurrent by Ceva's theorem. The medians intersect at the *centroid* of the triangle, P in Figure 6.3. The centroid is often denoted by G, as it is also known as the *center of gravity* of the triangle.

From Stewart's theorem, we have

$$m_a^2 = \frac{b^2 + c^2}{2} - \frac{a^2}{4}, \ m_b^2 = \frac{a^2 + c^2}{2} - \frac{b^2}{4}, \text{ and } m_c^2 = \frac{a^2 + b^2}{2} - \frac{c^2}{4}.$$

This is *Apollonius's theorem*, named for Apollonius of Perga (c. 262–190 BCE). It now follows that

$$m_a^2 - m_b^2 = \frac{3}{4}(b^2 - a^2),$$

so that if $a \le b \le c$, then $m_a \ge m_b \ge m_c$. Also, $m_a^2 + m_b^2 + m_c^2 = 3(a^2 + b^2 + c^2)/4$.

Three concurrent cevians partition a triangle into six smaller triangles. When the cevians are the medians, the triangles have equal area. To prove this, let AX, BY, CZ be the medians of ABC, and let x, y, z, u, v, w denote the areas of the six small triangles, as shown in Figure 6.6.

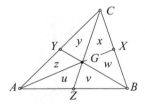

Figure 6.6.

Since a median divides a triangle into two triangles with equal areas, we have $u = v$, $w = x$, and $y = z$. Furthermore, $u + v + w = [ABC]/2 = v + w + x$, so that $u = x$. Similarly, $v = y$ and $w = z$, and thus $u = v = y = z = w = x$.

It now follows that the centroid trisects the medians. In Figure 6.6, $[ACG] = 2[CGX]$, and since the two triangles have the same altitude from C, $|AG| = 2|GX|$, and thus $|AX| = 3|GX|$. Similarly, $|BY| = 3|GY|$ and $|CZ| = 3|GZ|$.

Let P be a point inside triangle ABC, and draw the line segments AP, BP, and CP. If the triangles ABP, BCP, and ACP have equal areas, then P must be the centroid G of ABC. To see this, if we let X be the midpoint of BC, then the area of the quadrilateral $ABXP$ is $[ABC]/2$, and hence P lies on the median AX. Similarly P lies on the other two medians and hence P is the centroid.

The medians of a triangle always form a triangle, called the *median triangle*, with area three-fourths that of the original triangle, as seen in Figure 6.7 [Hungerbühler, 1999].

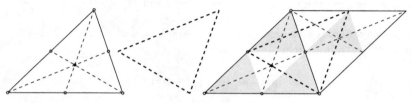

Figure 6.7.

6.3   Altitudes and the orthocenter

Cevians drawn perpendicular to the sides of a triangle are the *altitudes* of the triangle, as illustrated in Figure 6.8 for an acute triangle ABC.

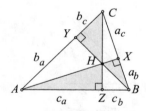

Figure 6.8.

Since triangles ABX and CBZ are similar, $a_b/c_b = c/a$, similarly we have $b_c/a_c = a/b$ and $c_a/b_a = b/c$, and hence

$$\frac{a_b}{a_c} \cdot \frac{b_c}{b_a} \cdot \frac{c_a}{c_b} = \frac{a_b}{c_b} \cdot \frac{b_c}{a_c} \cdot \frac{c_a}{b_a} = \frac{c}{a} \cdot \frac{a}{b} \cdot \frac{b}{c} = 1,$$

so by Ceva's theorem, the altitudes are concurrent (the proof is similar for obtuse triangles). The point of concurrency H is known as the *orthocenter* of the triangle (the Greek word *orthos* means straight or upright, as in perpendicular). The length of the altitudes are commonly denoted by $h_a = |AX|$, $h_b = |BY|$, and $h_c = |CZ|$ in Figure 6.8.

Since the area $[ABC]$ of the triangle is given by $ah_a/2$, $bh_b/2$, $ch_c/2$, as well as by Heron's formula $[ABC] = \sqrt{s(s-a)(s-b)(s-c)}$ where s denotes the semiperimeter $(a+b+c)/2$, it is easy to express the length of the altitudes as a function of the sides. For example, $h_a = 2\sqrt{s(s-a)(s-b)(s-c)}/a$.

The lengths of the altitudes can also be expressed by sines of the angles. Since $h_c = a \sin B = b \sin A$, we have $a/\sin A = b/\sin B$. With similar results for h_a and h_b, we have the *law of sines* for the triangle ABC:

$$\frac{a}{\sin A} = \frac{b}{\sin B} = \frac{c}{\sin C}. \tag{6.8}$$

In 1775, Giovanni Francesco Fagnano dei Toschi (1715–1797) posed the following problem: given an acute triangle, find the inscribed triangle of minimum perimeter. By *inscribed triangle* in a given triangle ABC, we mean a triangle XYZ such that each vertex X, Y, Z lies on a different side of ABC. Fagnano solved the problem using calculus, but we present a non-calculus solution using reflection and symmetry due to Lipót Fejér (1880–1959) [Kazarinoff, 1961]. The inscribed triangle with minimum perimeter is the *orthic triangle*—the triangle XYZ whose vertices are the feet of the altitudes from each of the vertices of ABC, as illustrated in Figure 6.9a.

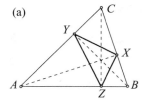 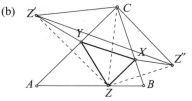

Figure 6.9.

To show that this is so, we find the locations for points X, Y, Z that minimize the perimeter of XYZ by choosing a point Z on side AB and reflecting it in sides AC and BC to locate points Z' and Z''. Thus the perimeter of XYZ is equal to $|Z'Y| + |YX| + |XZ''|$. The perimeter of XYZ will be a minimum whenever Z', Y, X, and Z'' lie on the same line. So for any Z, this gives the optimal location for X and Y. To find the optimal location for Z, note that triangle $Z'CZ''$ is isosceles, with $|CZ'| =$

$|CZ''| = |CZ|$ and vertex angle $\angle Z'CZ'' = 2\angle ACB$. Since the size of
the vertex angle of triangle $Z'CZ''$ does not depend on Z, the base $Z'Z''$
(the perimeter of XYZ) will be shortest when the legs are shortest, which
occurs when $|CZ|$ is a minimum, i.e., when CZ is perpendicular to AB.
Since triangle XYZ has minimum perimeter, whatever property Z has with
respect to C, X and Y have the same property with respect to A and B, as
we could have begun the proof by choosing and reflecting X or Y instead
of Z.

To conclude this section, see Figure 6.8 and note that triangle BHX
is similar to triangle AHY, hence $[BHX]/[AHY] = a_b^2/b_a^2$. Similarly
$[CHY]/[BHZ] = b_c^2/c_b^2$ and $[AHZ]/[CHX] = c_a^2/a_c^2$. Since the altitudes are concurrent cevians, (6.1) yields

$$\frac{[AHZ][BHX][CHY]}{[AHY][BHZ][CHX]} = \left(\frac{a_b}{a_c} \cdot \frac{b_c}{b_a} \cdot \frac{c_a}{c_b}\right)^2 = 1,$$

that is, the product of the areas of the white triangles in Figure 6.8 equals the
product of the areas of the gray triangles. Compare this result to the one for
the medians illustrated in Figure 6.6.

6.4   Angle-bisectors and the incenter

The three *angle-bisectors* of a triangle are the cevians bisecting its angles,
and are commonly denoted w_a, w_b, and w_c. We again use Ceva's theorem
to show they are concurrent. To do so, we need the

Angle-bisector theorem. *The bisector of an angle in a triangle divides
the side opposite the angle in the same ratio as the sides adjacent to the
angle.*

See Figure 6.10.

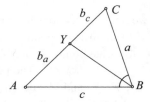

Figure 6.10.

Let BY bisect angle B. Since $\angle ABY = \angle CBY$ and $\angle AYB + \angle CYB = 180°$, we have

$$\frac{b_c}{a} = \frac{\sin \angle CBY}{\sin \angle CYB} = \frac{\sin \angle ABY}{\sin \angle AYB} = \frac{b_a}{c},$$

so $b_c/b_a = a/c$.

If AX and CZ are the angle bisectors of A and C, respectively, in Figure 6.3, we also have $a_b/a_c = c/b$ and $c_a/c_b = b/c$. Thus by Ceva's theorem, the angle bisectors are concurrent. The point of concurrency I (see Figure 6.11) is called the *incenter* of the triangle, as it is the center of the inscribed circle (the *incircle*), with radius r, the *inradius* of the triangle.

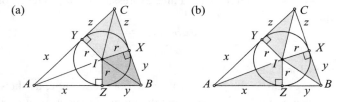

Figure 6.11.

Since the two small white right triangles in Figure 6.11a are congruent, as are the two light gray and the two dark gray triangles, we can label the segments of the sides as shown in Figure 6.11a, so that $a = y+z$, $b = z+x$, and $c = x+y$. Since $a+b+c = 2(x+y+z)$, the semiperimeter s equals $x+y+z$. The area $[ABC]$ of the triangle is the sum of the areas $[AIB]$, $[BIC]$, and $[CIA]$, and hence

$$[ABC] = ar/2 + br/2 + cr/2 = r(x+y+z) = rs.$$

Heron's formula $[ABC] = \sqrt{s(s-a)(s-b)(s-c)}$ can now be expressed as $[ABC] = \sqrt{sxyz}$, and hence $[ABC]rs = [ABC]^2 = sxyz$, or equivalently,

$$[ABC] = xyz/r. \tag{6.9}$$

Since the six right triangles in Figure 6.11b are congruent in pairs, we have

$$[AIZ] + [BIX] + [CIY] = r(x+y+z)/2 = [AIY] + [BIZ] + [CIX],$$

that is, the sum of the areas of the white triangles in Figure 6.11b equals the sum of the areas of the gray triangles. Compare this result to those for the medians and angle-bisectors in previous sections.

The lengths w_a, w_b, and w_c of the three angle-bisectors can be found using Stewarts' theorem. Since $c_a/c_b = b/c$, it follows that $c_a = bc/(a+b)$ and $c_b = ac/(a+b)$. Hence (6.5) yields

$$a^2 \cdot \frac{bc}{a+b} + b^2 \cdot \frac{ac}{a+b} = c\left(w_c^2 + \frac{abc^2}{(a+b)^2}\right),$$

which simplifies to

$$w_c^2 = ab\left(1 - \frac{c^2}{(a+b)^2}\right).$$

Since $(a+b)^2 - c^2 = 4s(s-c)$, a nice expression for w_c in terms of the semiperiemter s is

$$w_c = \frac{2\sqrt{ab}}{a+b}\sqrt{s(s-c)}$$

(and similarly for w_a and w_b).

If we draw cevians to the points of tangency of the incircle, as shown in Figure 6.12, they too are concurrent, and the point of concurrency is called the *Gergonne point,* after the French mathematician Joseph Diaz Gergonne (1771–1859). The concurrency follows from Ceva's theorem, noting that $c_a = b_a = x$, $a_b = c_b = y$, and $a_c = b_c = z$ in (6.1). See Figures 6.3 and 6.11.

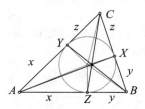

Figure 6.12.

6.5   Circumcircle and circumcenter

The vertices of a triangle determine a circle, called the *circumcircle* (for circumscribed circle) of the triangle. The triangle's *circumcenter* (the center of the circumcircle) is equidistant from the vertices, and hence lies on each of the three perpendicular bisectors of the sides. The radius R of the circumcircle is called the *circumradius* of the triangle.

Leonhard Euler (1707–1783) discovered that the circumcenter O, the orthocenter H, and the centroid G of a triangle determine a line (the *Euler line*), and that $|GH| = 2|GO|$. See Figure 6.13.

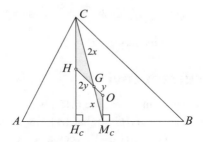

Figure 6.13.

Since CH is parallel to OM, the shaded triangles are similar and $|CG| = 2|GM_c|$, and thus $|GH| = 2|GO|$.

Analogous to $[ABC] = xyz/r$ in (6.9) relating the area $[ABC]$ to the inradius r and the lengths x, y, z in Figure 6.11 is a relationship between $[ABC]$, the circumradius R, and the side lengths: $[ABC] = abc/4R$. See Figure 6.14.

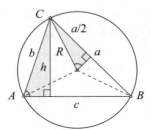

Figure 6.14.

The two shaded right triangles are similar, hence $h/b = (a/2)/R$, so that $h = ab/2R$. Thus $[ABC] = ch/2 = abc/4R$.

In the smaller shaded triangle, $a/2 = R \sin A$ and hence $a/\sin A = 2R$. Thus the common value of the fractions in the sine law (6.8) is $2R$:

$$\frac{a}{\sin A} = \frac{b}{\sin B} = \frac{c}{\sin C} = 2R. \tag{6.10}$$

Using the notation from Figure 6.11, the arithmetic mean-geometric mean inequality (see Section 2.2 or 4.2), and (6.9) now yields

$$4R[ABC] = abc = (x + y)(y + z)(z + x)$$
$$\geq 2\sqrt{xy} \cdot 2\sqrt{yz} \cdot 2\sqrt{zx} = 8xyz = 8r[ABC],$$

establishing *Euler's triangle inequality*: $R \geq 2r$.

6.6   Non-concurrent cevians

Cevians need not be concurrent. In Figure 6.15a we have drawn cevians to points X, Y, Z on the sides of ABC so that $|AY| = (1/3)|AC|$, $|BZ| = (1/3)|AB|$, and $|CX| = (1/3)|BC|$. The cevians are not concurrent but rather form a central triangle, sometimes called an *aliquot triangle*, shaded gray in Figure 6.15a. In Figure 6.15b we draw lines parallel to AX, BY, and CZ, to create triangles congruent to the central triangle, showing that the area of the aliquot triangle is $1/7$ of the area of the original triangle ABC [Johnston and Kennedy, 1993].

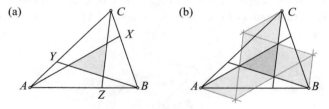

Figure 6.15.

The cevians AX, BY, CZ also form a triangle, with area $7/9$ of the area of ABC, as seen in Figure 6.16.

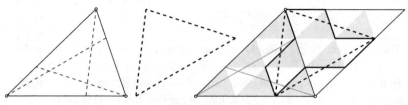

Figure 6.16.

If the points X, Y, and Z are chosen so that $|AY| = (1/n)|AC|$, $|BZ| = (1/n)|AB|$, and $|CX| = (1/n)|BC|$, then the area of the aliquot triangle is

$(n-2)^2/(n^2-n+1)$ of the area of ABC, and the area of the cevian triangle is $(n^2-n+1)/n^2$ of the area of ABC. See [Satterly, 1954, 1956] for details.

6.7 Ceva's theorem for circles

Suppose three chords AX, BY, CZ are concurrent at a point P inside an arbitrary circle, as illustrated in Figure 6.17. Then we have the same relationship for the six sides of the resulting inscribed hexagon that we had in (6.1):

$$\frac{|AZ|}{|ZB|} \cdot \frac{|BX|}{|XC|} \cdot \frac{|CY|}{|YA|} = 1.$$

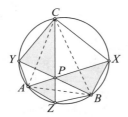

Figure 6.17.

To prove this [Hoehn, 1989], note that triangles APZ and CPX are similar, as are triangles BPX and APY, and triangles CPY and BPZ. Thus

$$\frac{|AZ|}{|XC|} = \frac{|AP|}{|XP|} = \frac{|PZ|}{|CP|}, \frac{|BX|}{|YA|} = \frac{|BP|}{|YP|} = \frac{|PX|}{|AP|}, \text{ and } \frac{|CY|}{|ZB|} = \frac{|CP|}{|ZP|} = \frac{|PY|}{|BP|}.$$

Therefore

$$\frac{|AZ|}{|ZB|} \cdot \frac{|BX|}{|XC|} \cdot \frac{|CY|}{|YA|} = \frac{|AZ|}{|XC|} \cdot \frac{|BX|}{|YA|} \cdot \frac{|CY|}{|ZB|}$$

$$= \left(\frac{|AP| \cdot |PZ|}{|XP| \cdot |CP|} \cdot \frac{|BP| \cdot |PX|}{|YP| \cdot |AP|} \cdot \frac{|CP| \cdot |PY|}{|ZP| \cdot |BP|}\right)^{1/2} = 1.$$

Furthermore, $[APZ]/[CPX] = |AZ|^2/|XC|^2$, $[BPX]/[APY] = |BX|^2/|YA|^2$, and $[CPY]/[BPZ] = |CY|^2/|ZB|^2$, hence

$$\frac{[APZ][BPX][CPY]}{[CPX][APY][BPZ]} = \left(\frac{|AZ|}{|ZB|} \cdot \frac{|BX|}{|XC|} \cdot \frac{|CY|}{|YA|}\right)^2 = 1.$$

That is, the product of the areas of the gray triangles in Figure 6.17 equals the product of the areas of the white triangles.

6.8   Challenges

6.1. Let m_a, m_b, and m_c be the medians and $s = (a + b + c)/2$ the semiperimeter of a triangle. Show that $3s/2 \le m_a + m_b + m_c \le 2s$.

6.2. Prove that if the sides of a triangle are in arithmetic progression, then the line through the centroid and the incenter is parallel to one of its sides.

6.3. Can you cut an arbitrary triangle into pieces so that the pieces can be rotated and translated (but not turned over) so as to form the mirror image of the given triangle? (It can be done in just two cuts!)

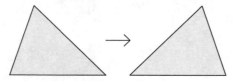

Figure 6.18.

6.4. Let ABC be a triangle whose longest side is AB, and let AX, BY, and CZ be cevians concurrent at P. Prove that $|PX| + |PY| + |PZ| < |AB|$.

6.5. Prove that the orthocenter H of triangle ABC in Figure 6.8 divides each altitude into two segments with the same product, i.e., $|AH| \cdot |HX| = |BH| \cdot |HY| = |CH| \cdot |HZ|$.

6.6. In triangle ABC, let X, Y, Z be the midpoints of BC, CA, and AB, respectively. Triangle XYZ is called the *medial triangle* of $\triangle ABC$. Prove that the orthocenter of $\triangle XYZ$ is the circumcenter of $\triangle ABC$.

6.7. Let ABC be a right triangle with C the right angle and CD the altitude to AB (see Figure 6.19). If CE is the median from C in $\triangle BCD$ and AC is extended its own length to F, show that FD and CE are perpendicular.

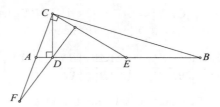

Figure 6.19.

6.8. Let AX, BY, and CZ be three cevians in triangle ABC, concurrent at P as shown in Figure 6.3. Prove that

(a) $\dfrac{|PX|}{|AX|} + \dfrac{|PY|}{|BY|} + \dfrac{|PZ|}{|CZ|} = 1$ and (b) $\dfrac{|PA|}{|AX|} + \dfrac{|PB|}{|BY|} + \dfrac{|PC|}{|CZ|} = 2.$

6.9. AX and BY are medians in triangle ABC, and P and Q are the midpoints of AY and BX, respectively (Figure 6.20). Prove that PQ is trisected by AX and BY—that is, $|PR| = |RS| = |SQ|$.

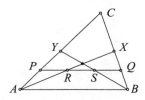

Figure 6.20.

6.10. Prove that the perimeter of every triangle of area $1/\pi$ exceeds 2.

CHAPTER **7**

The Right Triangle

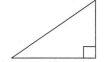

The sum of the square roots of any two sides of an isosceles triangle is equal to the square root of the remaining side.

The Scarecrow
The Wizard of Oz (1936)

The Pythagorean theorem, arguably the best-known theorem in geometry (in spite of the Scarecrow's misquote), is also the best-known property of right triangles. Right triangles have many other intriguing properties, and have figured prominently in each of the preceding chapters. So we now turn our attention to this important geometrical icon.

Unlike general triangles, the sides of a right triangle have special names. The sides adjacent the right angle are the *legs* (by analogy with the human body), and the third side is the *hypotenuse* (from the Greek *hypo-*, "under," and *teinein*, "to stretch," as the hypotenuse stretches from leg to leg).

In Figure 7.1 we see some common objects in the shape of right triangles that can be found around the home or the workplace: two drafting triangles, a sandwich box, and a corner table.

Figure 7.1.

Many other common objects have right triangular shapes, such as shelf brackets, some roof trusses, corner braces, etc.

In this chapter we will explore the use of right triangles in illustrating a variety of mathematical inequalities, investigate some of the special properties of various circles associated with right triangles, and examine Pythagorean triples (right triangles with integer sides). We will usually use the customary notation for triangles: A, B and C denote the angles (and ABC the triangle), and a, b, and c the sides opposite A, B, and C, respectively. When ABC is a right triangle often C will be the right angle and c the hypotenuse.

7.1   Right triangles and inequalities

In a right triangle the hypotenuse c is the longest side, and so we always have $a < c$ and $b < c$. In fact, virtually any inequality of the form $b \leq c$ (allowing for triangles with $a = 0$) can be illustrated with a right triangle by computing $a = \sqrt{c^2 - b^2}$ and using a and b as the legs and c the hypotenuse of a right triangle. Many inequalities for the various means of two positive numbers x and y can be illustrated this way. For example, if we set $c = (x + y)/2$, $b = \sqrt{xy}$, and $a = |x - y|/2$, we see the *arithmetic mean-geometric mean inequality* $(x + y)/2 \geq \sqrt{xy}$ in Figure 7.2.

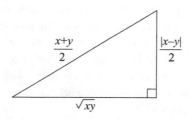

Figure 7.2.

In Table 7.1 we give the values of c, b, and a for the following inequalities given by $c \geq b$: 1) geometric mean-harmonic mean, 2) arithmetic mean-geometric mean, 3) root mean square-arithmetic mean, and 4) contraharmonic mean-root mean square. The illustrations all look like Figure 7.2 with the sides of the triangle relabeled. We have equality in each case if and only if $a = 0$, i.e., $x = y$.

Proposition 20 in Book I of Euclid's *Elements* reads "In any triangle the sum of any two sides is greater than the remaining one," that is, the three inequalities $a + b > c$, $b + c > a$, and $c + a > b$ all hold. When ABC is a right triangle with hypotenuse c, only $c \leq a + b$ (allowing for either a or

Table 7.1.

	c	b	a
1)	$\sqrt{xy}$	$\dfrac{2xy}{x+y}$	$\dfrac{\lvert x-y \rvert \sqrt{xy}}{x+y}$
2)	$\dfrac{x+y}{2}$	$\sqrt{xy}$	$\dfrac{\lvert x-y \rvert}{2}$
3)	$\sqrt{\dfrac{x^2+y^2}{2}}$	$\dfrac{x+y}{2}$	$\dfrac{\lvert x-y \rvert}{2}$
4)	$\dfrac{x^2+y^2}{x+y}$	$\sqrt{\dfrac{x^2+y^2}{2}}$	$\dfrac{\lvert x-y \rvert}{x+y}\sqrt{\dfrac{x^2+y^2}{2}}$

b to equal zero) is non-trivial. For example if we set $a = \sqrt{x}$, $b = \sqrt{y}$, and $c = \sqrt{x+y}$, we have $\sqrt{x+y} \le \sqrt{x} + \sqrt{y}$, as illustrated in Figure 7.3a. A function f that satisfies $f(x+y) \le f(x) + f(y)$ is called *subadditive*, and we have just shown that $f(x) = \sqrt{x}$ is subadditive on its domain.

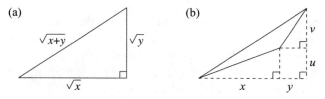

(a)   (b)

Figure 7.3.

For another example, see Figure 7.3b where we compute the length of the hypotenuse for each of three right triangles to establish the inequality: for positive numbers $x, y, u,$ and v,

$$\sqrt{(x+y)^2 + (u+v)^2} \le \sqrt{x^2 + u^2} + \sqrt{y^2 + v^2}.$$

The same procedure yields the familiar formula for the distance between two points in coordinate geometry. For additional examples, see Challenges 7.1, 7.2, and 7.7.

7.2 The incircle, circumcircle, and excircles

With every triangle we can associate five circles: the *incircle*, which lies inside the triangle and is tangent to each of the three sides; the *circumcircle*, which passes through the three vertices; and three *excircles*, each of which lies outside the triangle tangent to one side and to extensions of the other two sides. In Figure 7.4a we see the incircle with its *incenter* I and *inradius* r, and the circumcircle with its *circumcenter* O and *circumradius* R. In Figure 7.4b we see the three excircles with their *excenters* I_a, I_b, I_c and *exradii* r_a, r_b, r_c, the subscript indicating the triangle's side of tangency to the excircle. We have illustrated the circles for a right triangle, the case of primary interest, although all triangles possess the five circles, centers, and radii.

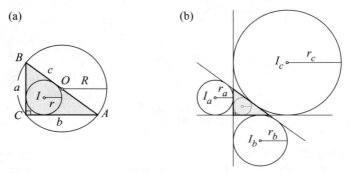

Figure 7.4.

There are many remarkable relationships among the five radii, the sides, the *semiperimeter* $s = (a+b+c)/2$ and the area K of the triangle. We begin with some relationships that hold for all triangles.

In Figure 7.5a, the area K of triangle ABC is equal to the sum of the areas of the three triangles AIB, BIC, and CIA, and hence

$$K = ar/2 + br/2 + cr/2 = r(a+b+c)/2 = rs.$$

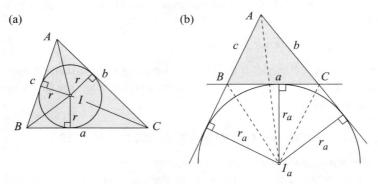

Figure 7.5.

Similarly, in Figure 7.5b K is equal to the sum of the areas of triangles AI_aC and AI_aB minus the area of BI_aC, and hence

$$K = \frac{1}{2}br_a + \frac{1}{2}cr_a - \frac{1}{2}ar_a = r_a\frac{b+c-a}{2} = r_a(s-a).$$

Similar results hold with r_b and r_c, and thus

$$K = rs = r_a(s-a) = r_b(s-b) = r_c(s-c). \tag{7.1}$$

As a consequence, we have

$$\frac{1}{r_a} + \frac{1}{r_b} + \frac{1}{r_c} = \frac{1}{r}.$$

Heron's formula $K = \sqrt{s(s-a)(s-b)(s-c)}$ and (7.1) now yield

$$K^2rr_ar_br_c = rs \cdot r_a(s-a) \cdot r_b(s-b) \cdot r_c(s-c) = K^4,$$

and thus $K = \sqrt{rr_ar_br_c}$. Finally, the circumradius R is related to K and a, b, c by $4KR = abc$ (for visual proofs of this result and Heron's formula, see [Alsina and Nelsen, 2006 and 2010]), and thus

$$
\begin{aligned}
4KR &= abc \\
&= s(s-b)(s-c) + s(s-a)(s-c) \\
&\quad + s(s-a)(s-b) - (s-a)(s-b)(s-c) \\
&= \frac{K^2}{s-a} + \frac{K^2}{s-b} + \frac{K^2}{s-c} - \frac{K^2}{s} \\
&= K(r_a + r_b + r_c - r).
\end{aligned}
$$

Hence the five radii of a triangle are related by $4R = r_a + r_b + r_c - r$.

Assume ABC is a right triangle with the right angle at C. Then $K = ab/2$ so that

$$
\begin{aligned}
r &= \frac{ab}{a+b+c}, r_a = \frac{ab}{b+c-a}, r_b \\
&= \frac{ab}{c+a-b}, \text{ and } r_c = \frac{ab}{a+b-c}. \tag{7.2}
\end{aligned}
$$

In Figure 7.6 we see that the hypotenuse $c = a + b - 2r$, and hence

$$r = \frac{a + b - c}{2} = s - c. \tag{7.3}$$

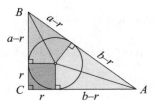

Figure 7.6.

Equating the expressions for r in (7.3) with the first one in (7.2) yields $(a + b - c)/2 = ab/(a + b + c)$, which simplifies to $a^2 + b^2 = c^2$—yet another proof of the Pythagorean theorem!

Similar expressions can be obtained for the exradii. For example

$$r_a = \frac{ab}{b - a + c} \cdot \frac{a - b + c}{a - b + c} = \frac{ab(a - b + c)}{c^2 - (a - b)^2}$$

$$= \frac{ab(a - b + c)}{c^2 - a^2 + 2ab - b^2} = \frac{ab(a - b + c)}{2ab} = \frac{a - b + c}{2} = s - b.$$

Similarly $r_b = s - a$ and $r_c = s$. Consequently, we have the following identities:

$$K = rs = s(s - c) = rr_c = (s - a)(s - b) = r_a r_b, r + r_a + r_b = r_c,$$

$$r + r_a + r_b + r_c = a + b + c,$$

$$r^2 + r_a^2 + r_b^2 + r_c^2 = a^2 + b^2 + c^2,$$

$$\text{and} \quad r_a r_b + r_b r_c + r_c r_a = s^2.$$

We can re-label the sides of ABC in Figure 7.6 to illustrate one of the above expressions for the area K. See Figure 7.7, where we show that K (which is also equal to the area of the white triangle in Figure 7.7a) equals the product of the lengths of the segments of the hypotenuse determined by the point of tangency of the inscribed circle.

For additional identities for the sides, area and radii of a right triangle, see [Long, 1983; Hansen, 2003; and Bell, 2006].

(a) 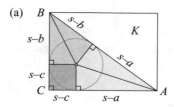 (b)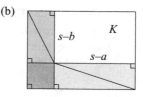

Figure 7.7.

We conclude this section with a characterization of right triangles in terms of r, R, and s: *ABC is a right triangle if and only if $s = r + 2R$.* Our proof is from [Blundon, 1963].

A triangle is a right triangle if and only if one of the three forms of the Pythagorean theorem holds, that is, if and only if

$$(a^2 + b^2 - c^2)(b^2 + c^2 - a^2)(c^2 + a^2 - b^2) = 0.$$

To establish the characterization we first need to express $ab + bc + ca$ and $a^2 + b^2 + c^2$ in terms of r, R and s for general triangles, which we now do:

$$\begin{aligned}
r^2 s = K^2/s &= (s-a)(s-b)(s-c) \\
&= s^3 - s^2(a+b+c) + s(ab+bc+ca) - abc \\
&= -s^3 + s(ab+bc+ca) - 4Rrs,
\end{aligned}$$

and hence $ab + bc + ca = s^2 + 4Rr + r^2$. Consequently

$$\begin{aligned}
a^2 + b^2 + c^2 &= 4s^2 - 2(ab+bc+ca) \\
&= 4s^2 - 2s^2 - 8Rr - 2r^2 = 2s^2 - 8Rr - 2r^2.
\end{aligned}$$

Thus

$$\begin{aligned}
0 &= (a^2 + b^2 - c^2)(b^2 + c^2 - a^2)(c^2 + a^2 - b^2) \\
&= [2(a^2b^2 + b^2c^2 + c^2a^2) - (a^4 + b^4 + c^4)](a^2 + b^2 + c^2) - 8(abc)^2 \\
&= 16[s(s-a)(s-b)(s-c)](2s^2 - 8Rr - 2r^2) - 8(4KR)^2 \\
&= 32K^2(s^2 - 4Rr - r^2 - 4R^2) \\
&= 32K^2(s + r + 2R)(s - r - 2R).
\end{aligned}$$

The only factor in the last line that is not always positive is $s - r - 2R$, and thus ABC is a right triangle if and only if $s = r + 2R$.

Characterizations of right triangles in terms of the exradii and the sides may be found in [Bell, 2006].

7.3   Right triangle cevians

Recall that in a triangle a cevian is a line segment that joins a vertex to a point on the opposite side. Examples of cevians are medians, altitudes, and angle-bisectors. In this section we present some special properties of right triangle cevians.

We have seen some of these special properties in earlier chapters. In Chapter 2 we saw that the bisector of the right angle in a right triangle also bisects the square on the hypotenuse, and in Chapter 4 that the median to the hypotenuse partitions a right triangle into two isosceles triangles.

The bisector of the right angle also bisects another angle, the angle between the altitude and median to the hypotenuse. In Figure 7.8a CD is the angle-bisector, CH the altitude, CM the median, and the claim is CD also bisects $\angle HCM$.

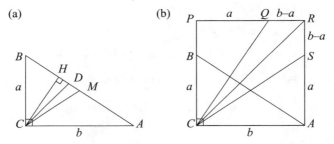

Figure 7.8.

Embed triangle ABC in a square whose sides are b, as shown in Figure 7.8b, and extend $CH, CD,$ and CM to intersect sides of the square at $Q, R,$ and S, respectively. Then triangles CPQ and CAS are congruent to triangle ABC, and hence triangles CQR and CRS are congruent, which establishes the result.

If we join the foot H of the altitude CH to the midpoints K and L of AC and BC, respectively, then the angle formed by the segments HK and HL is a right angle. See Figure 7.9.

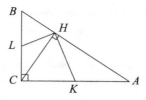

Figure 7.9.

Because HK and HL are medians of triangles ACH and BCH, respectively, triangles AKH and BLH are isosceles. Thus $\angle AHK = \angle A$ and $\angle BHL = \angle B$, and since $\angle A + \angle B = 90°$, $\angle KHL$ is a right angle.

7.4   A characterization of Pythagorean triples

Right triangles with integer sides, such as $(a, b, c) = (3, 4, 5)$ or $(5, 12, 13)$, are of particular interest. A triple (a, b, c) of positive integers such that $a^2 + b^2 = c^2$ is called a *Pythagorean triple*, and it is called *primitive* whenever a, b, and c have no common factors (such as in the two examples above).

(a)   (b)

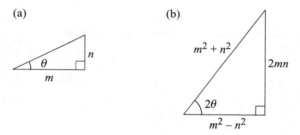

Figure 7.10.

The double angle formulas for the sine and cosine (see Section 4.6) yield an efficient way to generate Pythagorean triples [Houston, 1994]. Let m and n be integers with $m > n > 0$, and consider the right triangle with legs m and n, as illustrated in Figure 7.10a. If θ is the smaller acute angle, we have $\sin \theta = n/\sqrt{m^2 + n^2}$ and $\cos \theta = m/\sqrt{m^2 + n^2}$.

The double angle formulas yield $\sin 2\theta = 2mn/(m^2 + n^2)$ and $\cos 2\theta = (m^2 - n^2)/(m^2 + n^2)$, so that 2θ is an acute angle in a right triangle with integer sides, as illustrated in Figure 7.10b.

Not all Pythagorean triples can be obtained in this way. For example, $(9, 12, 15)$ cannot, since 15 is not the sum of two integral squares. But it can be shown that when m and n are relatively prime with one even and the other odd, a primitive Pythagorean triple results, and all primitive Pythagorean triples can be so generated.

For other methods to generate Pythagorean triples, see Challenge 7.4 and Section 13.3.

7.5 Some trigonometric identities and inequalities

Since the trigonometric functions are often defined in terms of ratios of sides in right triangles, right triangles can be used to illustrate a variety of trigonometric identities. In various sections and challenges in Chapters 2, 3, and 4 we saw right triangles used to illustrate the sum and difference formulas and double angle formulas for the sine, cosine, and tangent, and other identities. We continue with a few more.

Similar right triangles abound in Figure 7.11, yielding the familiar reciprocal and Pythagorean identities for the six trigonometric functions of a first quadrant angle θ. The figure also illustrates many less familiar identities, such as

$$\sec^2 \theta + \csc^2 \theta = (\tan \theta + \cot \theta)^2 \quad \text{and} \quad \tan \theta = \frac{\sec \theta - \cos \theta}{\sin \theta}.$$

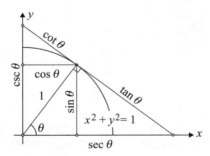

Figure 7.11.

The triple angle formulas for the sine and cosine can be computed from Figure 7.12, using a diagram based on the angle-trisection method of Archimedes [Dancer, 1937].

Computing $\sin \theta$ using the shaded triangle yields

$$\sin \theta = \frac{\sin 3\theta}{1 + 2\cos 2\theta} = \frac{\sin 3\theta}{3 - 4\sin^2 \theta}$$

and hence $\sin 3\theta = 3 \sin \theta - 4 \sin^3 \theta$. Similarly,

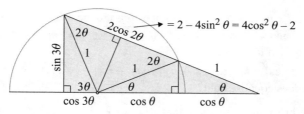

Figure 7.12.

$$\cos \theta = \frac{\cos 3\theta + 2\cos \theta}{1 + 2\cos 2\theta} = \frac{\cos 3\theta + 2\cos \theta}{4\cos^2 \theta - 1},$$

so that $\cos 3\theta = 4\cos^3 \theta - 3\cos \theta$. Challenge 7.11 is to use Figure 7.12 to compute the double angle formulas for the sine and cosine. For another way to use right triangles to obtain the triple angle formulas for the sine and cosine, see [Okuda, 2001].

We conclude with an inequality for the sum of tangents of n acute angles: if $\theta_k \geq 0$ for $k = 1, 2, \cdots, n$ and $\sum_1^n \theta_k < \pi/2$, then

$$\tan \left(\sum_1^n \theta_k \right) \geq \sum_1^n \tan \theta_k.$$

For a proof, see Figure 7.13 [Pratt, 2010].

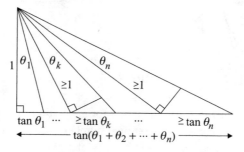

Figure 7.13.

7.6   Challenges

7.1. Use right triangles to illustrate

(a) the *Cauchy-Schwarz inequality* in two dimensions: for all real a, b, x, y, $|ax + by| \leq \sqrt{a^2 + b^2}\sqrt{x^2 + y^2}$ with equality if and only if $a/b = x/y$.

(b) *Aczél's inequality* in two dimensions: for real a, b, x, y such that $a^2 \geq b^2$ and $x^2 \geq y^2$, $\sqrt{a^2 - b^2}\sqrt{x^2 - y^2} \leq |ax - by|$ with equality if and only if $a/b = x/y$.

(c) for all real a, b, x, y, $|ax + by| + |ay - bx| \geq \sqrt{a^2 + b^2}\sqrt{x^2 + y^2}$. When does equality hold?

7.2. Use right triangles to show that if $x \geq y \geq 0$, then $\sqrt{3y^2 + x^2} \leq x + y \leq \sqrt{3x^2 + y^2}$.

7.3. (a) Prove that the vertices of a right triangle ABC lie on the sides of the triangle $I_a I_b I_c$ determined by the excenters of ABC. See Figure 7.14a.

(b) Prove that the common tangents to the excircles of the right triangle ABC indicated by the dashed lines in Figure 7.14b are parallel to one another and perpendicular to the line of the hypotenuse of ABC.

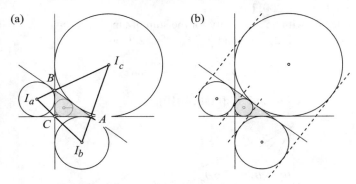

Figure 7.14.

7.4. In [Williamson, 1953] we find a rule for generating right triangles with rational sides: Take any two rational numbers whose product is 2 and add 2 to each. The results are the legs of right triangle with rational sides. For example, $(7/3)(6/7) = 2$, so 13/3 and 20/7 are sides of right triangle with a rational hypotenuse 109/21. Clearing fractions yields the Pythagorean triple (60, 91, 109). Is the rule valid?

7.5. (a) Prove that if each leg of a right triangle is rotated about its vertex on the hypotenuse so as to lie on the hypotenuse, then the legs overlap in a segment whose length is the diameter of the incircle. See Figure 7.15a.

(b) Prove that if the altitude CD to the hypotenuse AB of triangle ABC is drawn, then the sum of the inradii r, r_1, and r_2 of triangles

ABC, ACD, and BCD, respectively, equals the altitude of ABC. See Figure 7.15b.

(a)   (b)

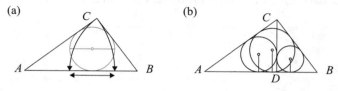

Figure 7.15.

7.6. If r, R, and K denote the inradius, circumradius, and area of a right triangle, prove that (a) $R + r \geq \sqrt{2K}$ and (b) $R/r \geq 1 + \sqrt{2}$. When does equality hold?

7.7. Show that if x and y are in $[0, 1]$, then $\sqrt{xy} + \sqrt{(1 - x)(1 - y)} \leq 1$ with equality if and only if $x = y$

7.8. In the right triangle ABC (with the right angle at C), let M be the midpoint of AC and N the point where the excircle with center I_a is tangent to BC, as shown in Figure 7.16. Show that the incenter I lies on MN.

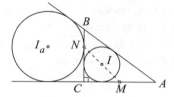

Figure 7.16.

7.9. Find the smallest right triangle with integer sides in which a square with integer sides can be inscribed so that all four vertices are on the triangle and one side of the square coincides with the hypotenuse. (This problem appeared in the November 1988 issue of the *College Mathematics Journal*.)

7.10. There are many ways to lace up one's shoes. In *The Shoelace Book* [Polster, 2006] there are diagrams for (a) the crisscross, (b) star, (c) zigzag, and (d) bowtie lacings for shoes with six pairs of eyelets.

If we assume uniform spacing of the eyelets horizontally and vertically, which lacing is the shortest? the longest? (Hint: no computations are necessary, since the eyelets are vertices of right triangles.)

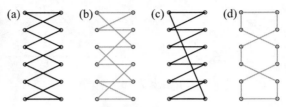

Figure 7.17.

7.11. Use Figure 7.12 to obtain the double angle formulas for the sine and cosine.

7.12. Show that

(a) $(\sin\alpha + \sin\beta)^2 + (\cos\alpha + \cos\beta)^2 = 2 + 2\cos(\alpha - \beta)$,

(b) $\tan((\alpha + \beta)/2) = (\sin\alpha + \sin\beta)/(\cos\alpha + \cos\beta)$.

(Hint: Consider a right triangle with legs $\sin\alpha + \sin\beta$ and $\cos\alpha + \cos\beta$).

7.13. A *mean* of two positive numbers is an expression symmetric in x and y whose value lies between $\min(x, y)$ and $\max(x, y)$. Hence $xy\sqrt{(x + y)/(x^3 + y^3)}$ and $\sqrt{(x^3 + y^3)/(x + y)}$ are means. How do they compare to the means in Section 7.1?

CHAPTER 8

Napoleon's Triangles

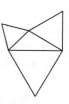

> *Un bon croquis vaut mieux qu'un long discours* (*A good sketch is better than a long speech*).
>
> Napoleon Bonaparte

In Chapter 1 we studied the bride's chair—a right triangle with squares erected on its sides. In this chapter we consider a similar icon—three equilateral triangles erected on the sides of a triangle. It figures prominently in the statements or proofs of a variety of results, including the one known as *Napoleon's theorem*.

Napoleon Bonaparte (1769–1821)

Did Napoleon prove the theorem with his name? No one knows, but we have given his name to the icon for this chapter. After Napoleon's theorem, we will examine a variety of other results such as the Fermat point of a triangle, Weitzenböck's inequality, Escher's theorem etc., all of which involve a triangle surrounded by triangles.

Triangles around triangles: Geodesic domes and spheres

The skeleton of a geodesic dome or sphere is a structure based on the great circles (geodesics) of a sphere. The circles intersect to form triangles around triangles, which have the structural rigidity of the triangle and give structural strength using a minimum amount of material. They have been built in many sizes, from small playground climbing structures to the large geodesic sphere at the NASA site in Florida, as seen in Figure 8.1.

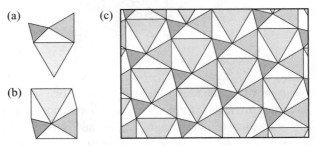

Figure 8.1.

8.1   Napoleon's theorem

The first surprise about our icon is that its component triangles tile the plane. In Figure 8.2a we see an arbitrary triangle in white surrounded by three gray equilateral triangles. In Figure 8.2b we see that the three gray triangles plus three copies of the white triangle form a hexagon, sometimes called a

Figure 8.2.

parahexagon since opposite edges are parallel and equal in length. And in Figure 8.2c we have a *tiling* of the plane by the hexagon (the *tile*), a placement of multiple copies of the tile that cover the plane with no gaps or overlaps.

Rotating the tiling in Figure 8.2c 120° clockwise or counterclockwise about the center of any equilateral triangle leaves the tiling unchanged. The rotational symmetry is sufficient to prove.

Napoleon's theorem. *The centers of the equilateral triangles constructed outwardly on the sides of a triangle are the vertices of another equilateral triangle.* See Figure 8.3a.

A portion of the tiling from Figure 8.2 is illustrated in Figure 8.3b. A 120° clockwise rotation of the tiling about P yields $|PQ| = |PS|$ and an angle of 120° between the segments. Similarly a 120° counterclockwise rotation of the tiling about R yields $|QR| = |RS|$ and an angle of 120° between those segments. Thus triangles PQR and PRS are congruent, so PR bisects both $\angle QPS$ and $\angle QRS$, and hence triangle PQR is equilateral. Triangle PQR is sometimes called the *outer Napoleon triangle* of $\triangle ABC$.

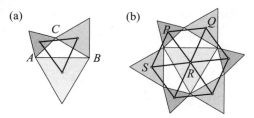

Figure 8.3.

8.2 Fermat's triangle problem

Our next surprise is the role played by the Napoleon triangles to solve the following problem, posed by Pierre de Fermat (1601–1665) for Evangelista Torricelli (1608–1647) to solve, which he did in several ways.

Find the point F in (or on) a triangle ABC such that the sum $|FA| + |FB| + |FC|$ is a minimum. (The point F is called the *Fermat point* of the triangle.) See Figure 8.4a.

When one of the angles measures 120° or more, its vertex is the Fermat point. So we will consider only triangles in which each angle measures less

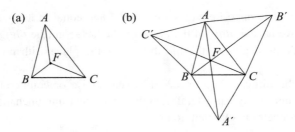

Figure 8.4.

than 120°. There is a simple way to locate the Fermat point of such a triangle. Construct equilateral triangles on the sides of ABC, and join each vertex of ABC to the exterior vertex of the opposite equilateral triangle. The three lines intersect at the Fermat point! See Figure 8.4b. That is why our icon is sometimes called *Torricelli's configuration*.

The proof we present was published by J. E. Hofmann in 1929, but it was not new at that time, having been found earlier by Tibor Gallai and others independently [Honsberger, 1973]. Take a point P inside ABC, connect it to the three vertices, and rotate the triangle ABP (shaded in Figure 8.5) 60° counterclockwise about point B as shown to triangle $C'BP'$, and join P' to P.

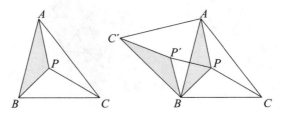

Figure 8.5.

By rotation $|AP| = |C'P'|$ and since triangle $BP'P$ is equilateral, $|BP| = |P'P|$, hence $|AP| + |BP| + |CP| = |C'P'| + |P'P| + |PC|$. Thus the sum $|AP| + |BP| + |CP|$ will be a minimum when P and P' lie on the straight line joining C' to C. There is nothing special about side AB and the new vertex C'—we could equally well have rotated BC or AC counterclockwise (or clockwise) about a vertex. Consequently P must also lie on $B'B$ and $A'A$ (not drawn in the figure), and the Fermat point F is P. In addition, each of the six angles at F measures 60° and the lines joining C to C', B to B', and A to A' (shown in Figure 8.4b, but not in 8.5) all have the same length: $|AP| + |BP| + |CP|$. Napoleon's theorem plays no role in the solution to Fermat's problem, perhaps because both Fermat and Torricelli died before Napoleon was born.

If we replace the equilateral triangles in Figure 8.4b with isosceles right triangles whose hypotenuses coincide with the sides of the given triangle, then the resulting cevians are concurrent and intersect at the Vecten point introduced in Chapter 1. The proof is immediate from Figure 1.11. Indeed, if we replace the equilateral triangles with a trio of similar isosceles triangles, the resulting cevians are concurrent, a result known as *Kiepert's theorem*. For a short proof, see [Rigby, 1991].

For another nice relationship between the Fermat point of a triangle ABC and the outer Napoleon triangle of ABC, see Challenge 12.10.

8.3 Area relationships among Napoleon's triangles

The *raison d'être* of the bride's chair is the area relationship among the three squares in the Pythagorean theorem. Similar results hold for Napoleon's triangles, and include the area of the central triangle as well.

Let ABC be a triangle with sides a, b, c (opposite vertices A, B, C, respectively), and let T denote its area. Also let T_s denote the area of an equilateral triangle with side s. See Figure 8.6.

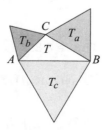

Figure 8.6.

As a consequence of Proposition 31 in Book VI of Euclid's *Elements* (a generalization of the Pythagorean theorem), $T_a + T_b = T_c$ when $\angle ACB = 90°$. However, we have surprisingly similar results when the angle is 60° or 120°.

(a) *If* $\angle ACB = 60°$, *then* $T + T_c = T_a + T_b$,

(b) *if* $\angle ACB = 120°$, *then* $T_c = T_a + T_b + T$.

For (a), we compute the area of an equilateral triangle with sides $a + b$ in two ways to conclude that $3T + T_c = 2T + T_a + T_b$, from which the result follows [Moran Cabre, 2003]. See Figure 8.7.

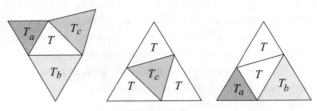

Figure 8.7.

For (b), we compute in two ways the area of an equiangular (120°) hexagon whose sides alternate in length between a and b to conclude that $3T + T_c = 4T + T_a + T_b$, which proves the result [Nelsen, 2004]. See Figure 8.8.

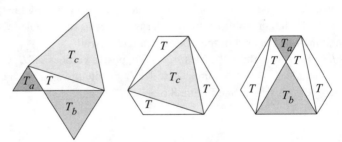

Figure 8.8.

In general, $T_c = T_a + T_b - \sqrt{3}\cot(\angle ACB) \cdot T$, and as a consequence we also have

(a) *If $\angle ACB = 30°$, then $3T + T_c = T_a + T_b$,*

(b) *if $\angle ACB = 150°$, then $T_c = T_a + T_b + 3T$*

(for visual proofs, see [Alsina and Nelsen, 2010]).

In addition to the relationships among T_a, T_b, T_c, and T for the special angles 30°, 60°, 90°, 120°, and 150°, we have *Weitzenböck's inequality*

$$T_a + T_b + T_c \geq 3T, \tag{8.1}$$

for all triangles ABC in Figure 8.6 with equality if and only if ABC is equilateral.

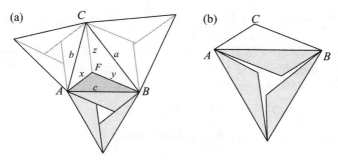

Figure 8.9.

To prove (8.1), we first consider the case where each angle of the triangle is less than $120°$. For such a triangle ABC, let x, y, and z denote the lengths of the line segments joining the Fermat point F to the vertices, as illustrated in Figure 8.9a, and note that the two acute angles in each triangle with a vertex at F sum to $60°$.

Hence the equilateral triangle with area T_c is the union of three triangles congruent to the dark gray shaded triangle with side lengths x, y, and c, and an equilateral triangle with side length $|x - y|$. Similar results are true for the triangles with areas T_a and T_b, and thus

$$T_a + T_b + T_c = 3T + T_{|x-y|} + T_{|y-z|} + T_{|z-x|}, \qquad (8.2)$$

which establishes (8.1) in this case since $T_{|x-y|}$, $T_{|y-z|}$, and $T_{|z-x|}$ are each nonnegative. We have equality in (8.1) if and only if $x = y = z$, so that the three triangles with a common vertex at F are congruent and hence $a = b = c$, i.e., triangle ABC is equilateral.

When one angle (say the one at C) measures $120°$ or more, then as illustrated in Figure 8.9b we have

$$T_a + T_b + T_c \geq T_c \geq 3T,$$

which completes the proof.

Weitzenböck's inequality can also be illustrated with Figure 8.3, which we used to prove Napoleon's theorem. Let K denote the area of the Napoleon triangle PQR whose vertices are the centers of the three equilateral triangles. Then Figure 8.3b shows that $6K = 3T + T_a + T_b + T_c$, or the regular hexagon in Figure 8.3b has the same area as the hexagonal tile in Figure 8.2b. Thus

$$K = \frac{1}{2}\left(T + \frac{T_a + T_b + T_c}{3}\right).$$

Since $6(K - T) = T_a + T_b + T_c - 3T$, the inequality $K \geq T$ is equivalent to Weitzenböck's inequality $T_a + T_b + T_c \geq 3T$.

The relationship in (8.2) is actually stronger than the Weitzenböck inequality, and enables us to prove the *Hadwiger-Finsler inequality*: If a, b, and c are the sides of a triangle, then

$$T_a + T_b + T_c \geq 3T + T_{|a-b|} + T_{|b-c|} + T_{|c-a|}.$$

For a proof, see [Alsina and Nelsen, 2008, 2009].

8.4 Escher's theorem

In the notebooks of the famous Dutch graphic artist Maurits Cornelis Escher (1898–1972), there are some nice results about hexagons that tile the plane. J. F. Rigby [Rigby, 1991] has arranged some of these results into

Escher's theorem. (i) *Let ABC be an equilateral triangle, and let E be any point, as in Figure 8.10. Let F be the point such that $|AF| = |AE|$ and $\angle FAE = 120°$. Let D be the point such that $|BD| = |BF|$ and $\angle DBF = 120°$. Then $|CE| = |CD|$ and $\angle ECD = 120°$. (ii) Congruent copies of the hexagon $AFBDCE$ tile the plane. (iii) The lines AD, BE, and CF (not drawn in Figure 8.10) are concurrent.*

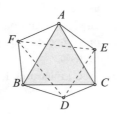

Figure 8.10.

To prove (i) we apply Napoleon's theorem to triangle DEF. Points A and B are the centroids of the equilateral triangles drawn on EF and FD, respectively. The centroids of all three equilateral triangles drawn on the sides of $\triangle DEF$ form an equilateral triangle by Napoleon's theorem, and $\triangle ABC$ is equilateral, so C must be the centroid of the equilateral triangle drawn on DE, and (i) follows (as Rigby notes, we must also assume that $\triangle ABC$ and $\triangle DEF$ have the same orientation, counterclockwise in the figure).

The plane tiling by congruent copies of hexagon $AFBDCE$ now follows immediately from Figure 8.2c by connecting the centroid of each equilateral triangle to its three vertices, as illustrated in Figure 8.11.

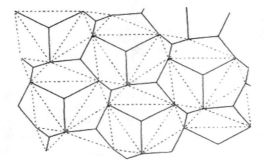

Figure 8.11.

Part (iii) of the theorem follows immediately form Kiepert's theorem, which we mentioned at the end of Section 8.2.

8.5 Challenges

8.1. In Figure 8.12 we see an incomplete Napoleon triangle configuration, with equilateral triangles APC and BQC drawn outwardly on two sides of $\triangle ABC$. Prove that the midpoint R of AB, the centroid G of $\triangle BQC$, and the vertex P determine a 30°–60°–90° triangle.

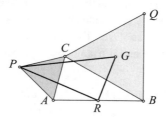

Figure 8.12.

8.2. Let $ABCD$ be a quadrilateral with $|AD| = |BC|$ and $\angle A + \angle B = 120°$, as shown in Figure 8.13a.

(a) Show that the midpoints X, Y, Z of the diagonals and side CD determine an equilateral triangle (see Figure 8.13b), and

(b) If equilateral triangle PCD is drawn outwardly on CD, then $\triangle PAB$ is also equilateral (see Figure 8.13c).

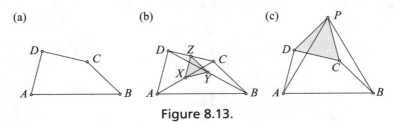

(a)   (b)   (c)

Figure 8.13.

8.3. Prove that if the sides of a triangle are trisected and equilateral triangles erected outwardly on the middle thirds of each side, then the vertices of the equilateral triangles form another equilateral triangle. See Figure 8.14.

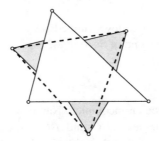

Figure 8.14.

8.4. In Section 1.2 we showed that in the Vecten configuration (three squares constructed on the sides of a triangle ABC). the three flank triangles each had the same area as ABC. Do triangles ABC exist for which the same is true when equilateral triangles are constructed on the sides of ABC?

8.5. Draw equilateral triangles ADE and CDF outwardly on adjacent sides of a rectangle $ABCD$, as shown in Figure 8.15. Show that triangle

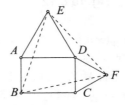

Figure 8.15.

BEF is equilateral, with side length equal to $\sqrt{3}/3$ times the side of the Napoleon triangle associated with the right triangle ACD.

8.6. Let ABC be an obtuse triangle with $\angle A > 90°$, and construct on its sides similar isosceles triangles $A'BC$, $AB'C$, and ABC' as shown in Figure 8.16, with vertex angles at A', B', and C' each equal to $2\angle A - 180°$. Let A'' be the reflection of A' in BC. Show that $AC'A''B'$ (shaded gray) is a parallelogram.

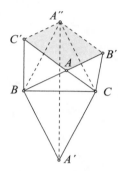

Figure 8.16.

(Hints: The base angles of the similar isosceles triangles are each equal to $180° - \angle A = \angle B + \angle C$, so that BAB' and CAC' are straight lines.)

8.7. The following problem appeared in the July-August 1923 issue of the *American Mathematical Monthly*: "Equilateral triangles are described on the sides of a right triangle. Dissect the triangles on the legs and reassemble the parts to form the triangle on the hypotenuse." Solve the problem.

CHAPTER 9

Arcs and Angles

*Every man of genius sees the world at a
different angle from his fellows.*
Havelock Ellis (1858–1939)

The icon for this chapter—an angle and a circular arc (or a complete circle)—is omnipresent in mathematics. It appears in a great many problems, theorems, constructions, and applications in geometry, astronomy, surveying, engineering, etc.

A multitude of historical and modern common objects exhibit angles and circles or circular arcs. Among these are, left to right in Figure 9.1, a pair of compasses, a navigational sextant, the clock on the Big Ben clock tower in London, and a Jeffersonian wind gauge for measuring wind speed and direction. Other such everyday objects include wagon wheels, scissors, protractors, and even a slice of pizza.

Sextant, p. 1932.

Figure 9.1.

We begin by exploring the relationship between arcs and angles in general. After presenting the notion of the power of a point, we prove Euler's triangle theorem, and discuss the Taylor circle of an acute triangle and the Monge circle of an ellipse.

9.1   Angles and angle measurement

Euclid defines an angle in the *Elements*, Book I, Definition 8, as "the inclination to one another of two lines in a plane which meet one another and do
not lie in a straight line" (Figure 9.2a). Today we say it is a figure defined
by a point (the vertex) and two rays emanating from that point; and the magnitude of the angle (the "inclination" in Euclid's definition) as the amount
of rotation needed for one ray to coincide with the other (Figure 9.2b). So
it is natural to measure the magnitude of the angle with an arc of a circle,
and this can be done with the classical Euclidean tools, the compass and
straightedge. While Euclid uses the right angle as the unit angle, the Babylonians introduced degrees for measuring angles, and Hypsicles of Alexandria
(circa 190–120 BCE) was the first Greek to divide the right angle into ninety
degrees.

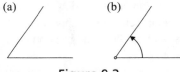

Figure 9.2.

Why are there 360° in a circle?

Historians of angle measurement credit the Babylonians for the division
of the circle into 360 parts, but they do not agree on the reason. Some
say it is because there are approximately 360 days in the year. Perhaps a
more plausible explanation comes from early Babylonian trigonometry,
based on the lengths of chords in a circle. A natural unit is a chord whose
length equals the radius of the circle, and consequently the Babylonian
unit angle was the angle of an equilateral triangle [Ball, 1960; Eves,
1969].

The Babylonians employed the sexagesimal (base-60) number system, so the unit angle was divided into 60 degrees (a term first used by
the Greeks), each degree was then divided into 60 minutes (from the
Latin *partes minutae primae*, "first small parts"), and each minute into
60 seconds (*partes minutae secundae*, "second small parts").

Euclid was able to prove many propositions in the *Elements* about angles
without needing to measure them. With some basic results about equal angles formed when a line crosses parallel lines, he proved that the sum of

the angles in any triangle equals two right angles (Book I, Proposition 32). A proof follows immediately from Figure 9.3 by drawing DE through C parallel to AB.

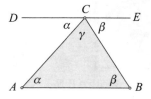

Figure 9.3.

For many applied problems angle measurement is essential, and an ancient simple tool for doing this is the protractor, a semicircle marked with degree measure (see Figure 9.4).

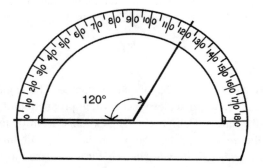

Figure 9.4.

The *sextant* (or *quadrant* or *octant*, depending on size of the arc used) is an instrument used by navigators and astronomers to measure the elevation of stars or the sun, as seen on the 1994 stamp from the Faröe Islands in Figure 9.5a. A simple one can be constructed from a protractor, a drinking straw, a piece of string, and a weight such as a washer. Turn the protractor upside down, tape or glue the straw to the now top edge of the protractor, and hang the weight on the string from the center of the protractor's semicircle, as shown in Figure 9.5b.

To measure the angle of elevation of an object, view it through the straw, and note the angle the string makes on the protractor (which will then have to be subtracted from 90°) to obtain the angle of elevation of the object.

(a)   (b)

   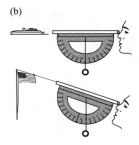

Figure 9.5.

Another unit for angle measurement is the *radian*. One radian is defined as the angle subtended by an arc whose length is the radius of the circle. Consequently 360° equals 2π radians. Radian measure is used in virtually all mathematics at the level of calculus and above.

There is a variety of other ways to measure angles, including the *grad* or gradian (360° equals 400 grads), the *mil*, an approximation to a milliradian (360° equals 6400 mils), the *point* (360° equals 32 points), etc. They are rarely encountered in mathematics.

Uncial calligraphy

In calligraphy each alphabet has its writing angle. The angle must be preserved in each letter to achieve a consistent production of thick and thin lines in the letters. Calligraphic angles range from 0° to 90°. For example, in the *uncial* alphabet the pen angle is 20°, in *half-uncial* it is 0°, in *carolingian* and *rotunda miniscule* it is 30° (the most common angle), *sharpened italic* requires 45°, and *gothic script* 90°. In the figure we see some Roman uncial script from the 5th century.

Figure 9.6.

9.2 Angles intersecting circles

In the previous section we saw how an arc of a circle centered at the angle vertex measures the size of the angle. We now extend this idea to angles and circles where the center of the circle and the vertex of the angle do not coincide.

Given a circle and an angle, we say that the angle is *central* if its vertex is the center of the circle, and *inscribed* if its vertex lies on the circle and its rays intersect the circle at two other points on the circle. We first show that *the measure of an inscribed angle is one-half the measure of the central angle subtending the same arc on the circle*. In Section 4.1 we saw the special case when one side of the angle is a diameter of the circle. This yields the result for the general case in Figure 9.7.

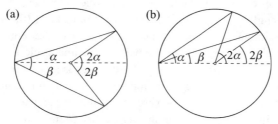

Figure 9.7.

This basic result has many useful consequences, including the following two:

(i) A convex quadrilateral may be inscribed in a circle if and only if its opposite angles are supplementary (i.e., sum to 180°).

(ii) Given a segment QR and an angle α, $0 < \alpha < 180°$, the set of points P such that $\angle QPR = \alpha$ is an arc of a circle.

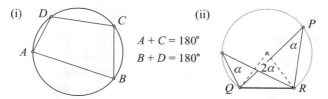

Figure 9.8.

The case $\alpha = 90°$ of (ii) is precisely Thales' triangle theorem from Section 4.1.

A *semi-inscribed* angle has its vertex on the circle with one side tangent to the circle, as illustrated in Figure 9.9. Once again, the measure of the angle is one-half the measure of the central angle.

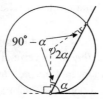

Figure 9.9.

At this point radian measure becomes simpler than degree measure, since the length of a circular arc is the radius of the circle times the radian measure of the central angle. If we use a circle with radius equal to 1 (the *unit circle*), then it is even simpler: arc length and radian measure of the central angle are the same. In the rest of this section we will use the unit circle.

We now consider an *interior angle* whose vertex is a point inside the circle, where for convenience we use a unit circle. Its radian measure is arithmetic mean of the lengths of the two arcs of the unit circle determined by the angle and its equal vertical angle at the vertex. See Figure 9.10a.

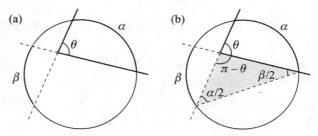

Figure 9.10.

Summing the measures of the angles in the gray triangle in Figure 9.10b yields $(\pi - \theta) + \alpha/2 + \beta/2 = \pi$, so that $\theta = (\alpha + \beta)/2$.

When the vertex is outside the unit circle (and both sides of the angle intersect the circle), we call the angle an *exterior angle*. Its measure is one-half the difference of the lengths of the two arcs, as shown in Figure 9.11a.

Summing the measures of the angles in the gray triangle in Figure 9.11b yields $\theta + \beta/2 + (\pi - \alpha/2) = \pi$, so that $\theta = (\alpha - \beta)/2$.

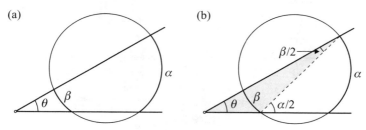

Figure 9.11.

9.3 The power of a point

Given a point P in the same plane as a circle C, consider the points of intersection Q and R of an arbitrary line through P and C. The *power of P with respect to C* is defined as the product $|PQ| \cdot |PR|$ [Andreescu and Gelca, 2000]. It is independent of the choice of the line QR. To see this, consider another line intersecting C at S and T, as shown in Figure 9.12 for both the case where (a) P is outside C and (b) P is inside C.

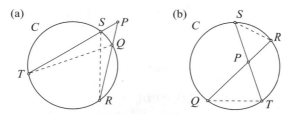

Figure 9.12.

In both cases, triangles PQT and PSR are similar, and hence $|PS|/|PQ| = |PR|/|PT|$, so that $|PS| \cdot |PT| = |PQ| \cdot |PR|$. Since the power of the point P is independent of the choice of the line through Q and R, we can choose the line through the center O of C, as shown in Figure 9.13a when P is outside C. If we let d be the distance between P and O and r the radius of C, then $|PQ| \cdot |PR| = (d-r)(d+r) = d^2 - r^2$. This is the square of the length $|PT|$ of the tangent line to C through P.

When P is inside C, we have $|PQ| \cdot |PR| = (r-d)(r+d) = r^2 - d^2$, and this is the square of half the length of the chord through P perpendicular to the diameter through P, as shown in Figure 9.13b. When P is on C, the power of P with respect to C is zero. Many authors *define* the power of a point as $d^2 - r^2$, so that power is negative when P is inside the circle. To

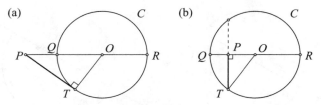

Figure 9.13.

make this definition consistent with ours requires signed distances between points.

If $ABCD$ is a convex quadrilateral that is not a rectangle, then at least one pair of opposite sides, say AB and CD, are not parallel. If AB and CD are extended to intersect at P and $|PA| \cdot |PB| = |PC| \cdot |PD|$, then $ABCD$ is a *cyclic quadrilateral*, that is, $ABCD$ has a circumcircle, as shown in Figure 9.14.

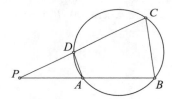

Figure 9.14.

To see this, suppose that $|PA| \cdot |PB| = |PC| \cdot |PD|$ and C is not on the circle determined by A, B, and D. Extend PD until it intersects the circle at C', different from C. The power of P with respect to this circle is $|PA| \cdot |PB| = |PC'| \cdot |PD|$, and hence $|PC| = |PC'|$, or $C = C'$, a contradiction.

For another example of the use of power of a point, consider the following problem. Suppose h_a, h_b, and h_c are computed as the altitudes to the sides a, b, and c in a triangle ABC. Is it possible to reconstruct ABC knowing only h_a, h_b, and h_c? The answer is yes, as we now show using the power of a point [Esteban, 2004].

From a point P extend three rays, and mark points Q, R, S on the rays where $|PQ| = h_a$, $|PR| = h_b$, and $|PS| = h_c$. See Figure 9.15. Let C be the circumcircle of triangle QRS, and let X, Y, and Z denote the second intersections of the rays with C.

By the power of P with respect to C, we have $|PX| h_a = |PY| h_b = |PZ| h_c$. However, the sides a, b, c of the triangle satisfy $a h_a = b h_b = c h_c$

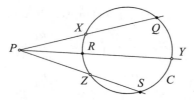

Figure 9.15.

(since each product is twice the area of ABC). Hence a triangle T with sides $|PX|$, $|PY|$, and $|PZ|$ is similar to ABC. Finding the altitude to side $|PX|$ and comparing it to T enables us to scale T to recover ABC.

9.4 Euler's triangle theorem

To illustrate the usefulness of the power of a point, we prove

Euler's triangle theorem. *Let I be the incenter and O the circumcenter of triangle ABC, with inradius r and circumradius R. See Figure 9.16. If d is the distance between O and I, then*

$$d^2 = |OI|^2 = R(R - 2r).$$

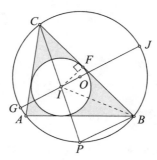

Figure 9.16.

Bisect $\angle C$ and extend the bisector through I to meet the circumcircle at P. Draw IO, and extend it to meet the circumcircle at G and J. The power of I with respect to the circumcircle is $|CI| \cdot |IP| = |GI| \cdot |IJ| = (R-d)(R+d) = R^2 - d^2$. Now $\angle ABP = \angle ACP = (1/2)\angle C$, and since IB bisects $\angle B$, $\angle IBP = (1/2)\angle B + (1/2)\angle C$. Applying the exterior angle theorem to $\triangle BIC$ yields $\angle PIB = (1/2)\angle B + (1/2)\angle C$, hence $\triangle IBP$ is isosceles and so $|IP| = |BP|$.

In $\triangle CIF$, $r = |CI| \sin(\angle ICF)$; and in $\triangle CPB$, we have

$$\frac{|PB|}{\sin(\angle PCB)} = \frac{|BC|}{\sin(\angle CPB)} = \frac{|BC|}{\sin(\angle A)} = 2R,$$

where the last step is from (7.10). Since $\angle PCB = \angle ICF$ and $|IP| = |BP|$, $2Rr = |CI| \cdot |BP| = R^2 - d^2$ and thus $d^2 = R^2 - 2Rr = R(R - 2r)$ as claimed.

The power of the incenter with respect to the circumcircle is $2Rr$.

Since squares are never negative, Euler's triangle theorem yields as a corollary *Euler's triangle inequality* (see Section 6.5) for any triangle with inradius r and circumradius R: $R \geq 2r$.

9.5   The Taylor circle

In an acute triangle ABC, draw the altitudes AX, BY, CZ, as shown in Figure 9.17. From the foot of each, draw segments perpendicular to the other two sides (the dashed lines in the figure). The resulting points (X_b, X_c, Y_c, Y_a, Z_a, and Z_b in the figure) lie on a circle, called the *Taylor circle* after H. M. Taylor (1842–1927), who discussed it in 1882 [Bogomolny, 2010].

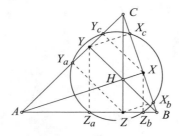

Figure 9.17.

To show that the six points are concyclic, we exploit the abundance of right triangles in the figure to show first that certain subsets of four of the six points are the vertices of cyclic quadrilaterals.

Since $\triangle AHZ$ is similar to $\triangle AXZ_b$, we have $|AZ|/|AZ_b| = |AH|/|AX|$, and since $\triangle AHY$ is similar to $\triangle AXY_c$, we have $|AH|/|AX| = |AY|/|AY_c|$, consequently $|AZ|/|AZ_b| = |AY|/|AY_c|$. Since $\triangle AY_aZ$ is similar to $\triangle AYZ_a$, we also have $|AY_a|/|AZ| = |AZ_a|/|AY|$, so that

$$\frac{|AY_a|}{|AZ|} \cdot \frac{|AZ|}{|AZ_b|} = \frac{|AZ_a|}{|AY|} \cdot \frac{|AY|}{|AY_c|},$$

which simplifies to $|AY_a| \cdot |AY_c| = |AZ_b| \cdot |AZ_a|$. Thus the four points Y_c, Y_a, Z_a, and Z_b lie on a circle, which we call C_A, and the previous expression gives the power of A with respect to C_1. Similarly the four points X_b, X_c, Z_a, and Z_b lie on a circle C_2, and the four points X_b, X_c, Y_c, and Y_a lie on a circle C_3.

We now need to show that the circles C_1, C_2, and C_3 coincide. If two coincide, then all three do, so assume that the circles are different. To reach a contradiction, we need the notion of the *radical axis* for a pair of circles, which is the locus of points having equal power with respect to the two circles [Andreescu and Gelca, 2000]. For intersecting circles, it is the line through the two points of intersection (since they have equal power, i.e., 0, with respect to both circles). Hence AB is the radical axis of C_1 and C_2, BC is the radial axis of C_2 and C_3, and AC is the radical axis of C_1 and C_3. Hence the vertices have equal power with respect to all three circles, which implies that vertex A, for example, lies on all three sides of the triangle, a contradiction.

9.6   The Monge circle of an ellipse

Sometimes circles appear unexpectedly as solutions to geometry problems. Such is the case in the following problem: What is the locus of points of the intersections of two perpendicular tangents to an ellipse? The locus is sometimes called the *orthoptic curve* of the ellipse, as it is the locus of points in the plane from which the ellipse is seen at a right angle. The French mathematician Gaspard Monge (1746–1818) investigated questions such as this one, and the locus—a circle centered at the center of the ellipse—is sometimes called the *Monge circle* or *director circle* of the ellipse. See Figure 9.18.

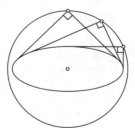

Figure 9.18.

The proof [Tanner and Allen, 1898] is surprisingly simple. If a line is tangent to the ellipse $(x^2/a^2) + (y^2/b^2) = 1$, then substituting $y = mx + k$ into the equation of the ellipse yields a quadratic in x with a single solution, from which it follows that $k^2 = a^2m^2 + b^2$. Hence the two lines tangent to the ellipse with slope m are

$$y - mx = \pm\sqrt{a^2m^2 + b^2}. \tag{9.1}$$

It now follows that the equations of the tangent lines perpendicular to (9.1) are $y + (1/m)x = \pm\sqrt{(a^2/m^2) + b^2}$, or equivalently

$$my + x = \pm\sqrt{a^2 + b^2m^2}. \tag{9.2}$$

Squaring (9.1) and (9.2) and adding yields $(m^2 + 1)x^2 + (m^2 + 1)y^2 = (m^2 + 1)(a^2 + b^2)$, and hence the locus of the intersections is the circle $x^2 + y^2 = a^2 + b^2$.

9.7   Challenges

9.1. Let ABC be a right triangle with C the right angle, and draw a circle with center B and radius $a = |BC|$, a shown in Figure 9.19. Prove the Pythagorean theorem by finding the power of A with respect to the circle.

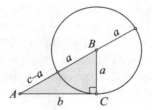

Figure 9.19.

9.2. Let P be a point inside a circle such that there exist three distinct chords through P of equal length. Prove that P is the center of the circle.

9.3. In Figure 9.17, show that $\{A, X, Z_b, Y_c\}$, $\{B, Y, X_c, Z_a\}$, and $\{C, Z, X_b, Y_a\}$ are sets of concyclic points.

9.4. In Figure 9.17, show that $\{X, Y, X_c, Y_c\}$, $\{Y, Z, Y_a, Z_a\}$, and $\{Z, X, Z_b, X_b\}$ are sets of concyclic points.

9.5. The W. M. Keck Observatory is located at the summit of the Mauna Kea volcano on the island of Hawaii (see Figure 9.20). Find (a) the power of the point of the observatory with respect to a great circle of the earth at sea level, and (b) the distance to the horizon from the observatory. (Hint: The observatory is about 4.2 km above sea level, and the mean radius of the earth is about 6378 km.)

Figure 9.20.

9.6. Use Figure 9.21 (part (a) for the acute case, and part (b) for the obtuse case) and powers of points to derive the law of cosines. (Hint: in both cases $CBZY$ and $ACXZ$ are cyclic quadrilaterals.)

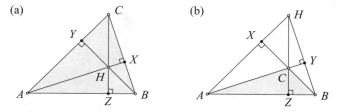

Figure 9.21.

9.7. In a circle with diameter AB, consider a chord CD parallel to AB, as shown in Figure 9.22. If x and y are the measures of $\angle ACD$ and $\angle ADC$, respectively, show that $x - y = 90°$.

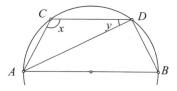

Figure 9.22.

Figure 2-4 (caption)

Figure 2-5

Figure 2-6

CHAPTER **10**

Polygons with Circles

> *"...I've something more important here. Something more lasting, you know. Something which will really endure"*
> *"What's that?"*
> *"Mind! Don't spoil my circles! It's the method of calculating the area of a segment of a circle."*
>
> Karel Čapek
> *The Death of Archimedes*

Euclid devoted Book IV of the *Elements* to propositions concerning inscribing and circumscribing polygons in or about circles, and circles in or about polygons. They have had a profound impact on geometry. For example, Archimedes, in *Measurement of the Circle*, was able to show that π is approximately 22/7 by inscribing and circumscribing regular polygons with 96 sides and computing the ratios of the perimeters of polygons to their diameters.

In Leonardo da Vinci's *Vitruvian Man*, seen in Figure 10.1 from his 15th century drawing and on a one euro coin from Italy, the artist is comparing the proportions of the human body to a circumscribed square and circle.

Figure 10.1.

Circles with polygons—especially squares—are common motifs in art and everyday objects. In Figure 10.2 we see Wassily Kandinsky's 1913 painting

Color Study: Squares with Concentric Rings, and similar designs on a shower curtain and a rug, which also appear in ceramics, jewelry, and corporate logos.

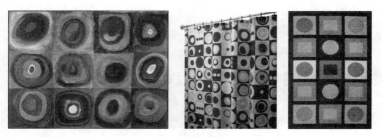

Figure 10.2.

Every triangle has an inscribed circle and a circumscribed circle, and we discussed their properties in Chapters 6 and 7. So we begin our explorations in this chapter with quadrilaterals—cyclic, tangential, and bicentric. For cyclic quadrilaterals we present Ptolemy's theorem, discuss the anticenter, and prove the Japanese theorem, a lovely result from the sangaku tradition. We also present Fuss's theorem for bicentric quadrilaterals and the butterfly theorem for a self-intersecting quadrilateral inscribed in a circle.

10.1   Cyclic quadrilaterals

A *cyclic quadrilateral* is one whose vertices lie on a circle. As we observed in Section 9.2, the angles at opposite vertices of a cyclic quadrilateral are supplementary. There are many lovely results about a cyclic quadrilateral Q with sides a, b, c, d, diagonals p and q, and circumradius R. We begin with Ptolemy's theorem for cyclic quadrilaterals. It is generally credited to Claudius Ptolemy of Alexandria (circa 85–165). There are many proofs of Ptolemy's theorem, but perhaps the nicest is the one given by Ptolemy himself in the *Almagest*, which we now present.

Ptolemy's theorem. *In a cyclic quadrilateral Q, the product of the diagonals equals the sum of the products of the two pairs of opposite sides, i.e., if Q has sides a, b, c, d (in that order) and diagonals p and q, then $pq = ac + bd$.*

Figure 10.3a illustrates a cyclic quadrilateral $ABCD$, and in Figure 10.3b we choose the point E on the diagonal BD and draw CE so that $\angle BCA = \angle DCE$. Let $|BE| = x$ and $|ED| = y$, with $x + y = q$.

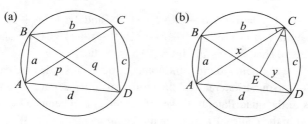

Figure 10.3.

The angles marked $\angle$ in Figure 10.4a subtend the same arc, hence they are equal and the shaded triangles are similar. Thus $a/p = y/c$, or $ac = py$. Similarly in Figure 10.4b the shaded triangles are similar, and $d/p = x/b$, or $bd = px$. Thus $ac + bd = p(x + y) = pq$.

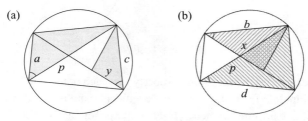

Figure 10.4.

Ptolemy's theorem can be strengthened to *Ptolemy's inequality* for convex quadrilaterals: *If Q is a convex quadrilateral with sides a, b, c, d (in that order) and diagonals p and q, then $pq \leq ac + bd$, with equality if and only if Q is cyclic.* See [Alsina and Nelsen, 2009] for a proof.

Within any convex quadrilateral is a cyclic one. To find it, bisect the angles and the four angle bisectors form a cyclic quadrilateral. See Figure 10.5.

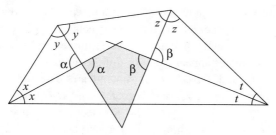

Figure 10.5.

Since $\alpha + x + y = 180°$, $\beta + z + t = 180°$, and $2x + 2y + 2z + 2t = 360°$, we have $\alpha + \beta = (180° - x - y) + (180° - z + t) = 180°$, and hence the quadrilateral shaded gray in Figure 10.5 is cyclic.

The two line segments joining midpoints of opposites sides are called the *bimedians* of the quadrilateral, and the bimedians intersect at the centroid of the vertices of the quadrilateral. If we reflect the circumcenter O of a cyclic quadrilateral in the centroid P, we obtain the *anticenter* O', as illustrated in Figure 10.6a.

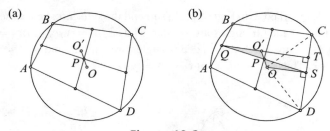

Figure 10.6.

The anticenter has the surprising property that each of the four lines drawn from the midpoint of one side through the anticenter are perpendicular to the opposite side, e.g., line segment $QO'T$ is perpendicular to CD in Figure 10.6b. This follows from the fact that triangles POS and $PO'Q$ are congruent, and triangle OCD is isosceles since OC and OD are circumradii. Hence OS is both a median and altitude in triangle OCD, and since OS is parallel to QO', $QO'T$ is perpendicular to CD.

When a cyclic quadrilateral has perpendicular diagonals, the diagonals intersect at the anticenter. There are many such—in a Cartesian coordinate system, all quadrilaterals whose vertices are the intersections of the coordinate axes with a circle containing the origin are in this class. Let Q be the intersection of the perpendicular diagonals of a cyclic quadrilateral, as shown in Figure 10.7. To show that Q is the anticenter, we show that a line passing through Q perpendicular to one side bisects the opposite side. In the figure we have drawn such a line and marked complementary angles x and y and where these angles appear in the rest of the figure.

Because $\angle BAC$ and $\angle BDC$ both subtend arc BC and $\angle ABD$ and $\angle ACD$ both subtend arc AD, triangles ARQ and BRQ are both isosceles, and thus $|AR| = |QR| = |BR|$. So R bisects AB. Since the same is true for the other three lines through Q perpendicular to a side, Q is the anticenter of $ABCD$.

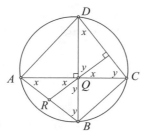

Figure 10.7.

There are many nice formulas and inequalities for the diagonals, area, and circumradius of cyclic quadrilaterals. For example, if a, b, c, d are the sides and K the area of a cyclic quadrilateral, then *Brahmagupta's formula* states that $K = \sqrt{(s-a)(s-b)(s-c)(s-d)}$, where $s = (a+b+c+d)/2$ is the semiperimeter of the quadrilateral. For a proof of Brahmagupta's formula and related results, see [Alsina and Nelsen, 2009, 2010].

10.2 Sangaku and Carnot's theorem

Sangaku (literally "mathematical tablet") are Japanese geometry theorems that were often written on wooden tablets during the Edo period (1603–1867) and hung on Buddhist temples and Shinto shrines as offerings. Figure 10.8 shows a sangaku with diagrams and text for five theorems.

Figure 10.8.

The following sangaku theorem is from about 1800, and is often called

The Japanese theorem. *If a polygon is inscribed in a circle and partitioned into triangles by diagonals, then the sum of the inradii of the resulting*

triangles is a constant independent of the particular triangulation of the polygon. See Figure 10.9 for an example showing two different triangulations of a cyclic pentagon.

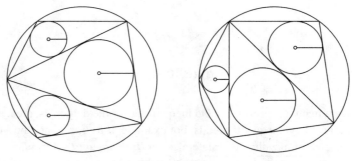

Figure 10.9.

Traditionally sangaku contained the text of the problem and a diagram, but not the proof. Our proof of the Japanese theorem uses *Carnot's theorem*, due to Lazare Nicolas Marguérite Carnot (1753–1823). For an acute triangle ABC, Carnot's theorem states that *the sum of the distances x, y, z from the circumcenter to the sides is equal to the sum of the inradius r and the circumradius R*, i.e., $x + y + z = r + R$. See Figure 10.10a.

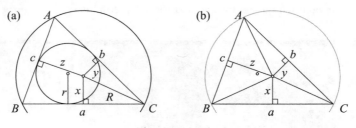

Figure 10.10.

To prove Carnot's theorem we relate r and R to x, y, z and the sides a, b, c. From Sections 6.5 and 7.2 we know that the area K of $\triangle ABC$ satisfies $2K = r(a + b + c)$. From Figure 10.10b we also have $2K = ax + by + cz$, and hence $ax + by + cz = r(a + b + c)$.

As noted following Figure 6.13, the angles at the circumcenter of $\triangle ABC$ in Figure 10.11a marked β and γ are equal to the anglesat vertices B and C.

Hence scaled versions of $\triangle ABC$ and the two smaller gray triangles in Figure 10.11a form a rectangle as shown in Figure 10.11b.

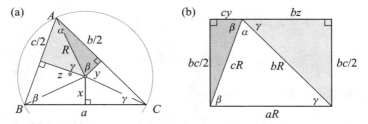

Figure 10.11.

Thus $cy + bz = aR$, and similarly $az + cx = bR$ and $bx + ay = cR$. Hence

$$(a + b + c)(x + y + z) = (ax + by + cz) + (cy + bz) + (az + cx)$$
$$+(bx + ay) = r(a + b + c) + (a + b + c)R = (a + b + c)(r + R),$$

so $x + y + z = r + R$ as claimed.

For an obtuse triangle ABC, we have the situation illustrated in Figure 10.12. In this case the circumcenter lies outside the triangle, so one of the perpendicular line segments drawn to the sides lies completely outside $\triangle ABC$, say x. Carnot's theorem in this case states that $y + z - x = r + R$.

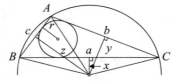

Figure 10.12.

In this case computing the area of $\triangle ABC$ yields $2K = by + cz - ax$ so that $by + cz - ax = r(a + b + c)$. With a figure similar to Figure 10.11b we again have $cy + bz = aR$. With angles labeled as is Figure 10.13a, we have $\beta + \gamma + \delta = \pi/2$ and hence we can construct the rectangle in Figure 10.13b with scaled copies of $\triangle ABC$ and the gray shaded triangles in Figure 10.13a to obtain $az - cx = bR$.

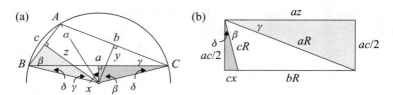

Figure 10.13.

Similarly, $ay - bx = cR$ and thus

$$(a + b + c)(y + z - x) = (by + cz - ax) + (cy + bz) + (az - cx)$$
$$+ (ay - bx) = r(a + b + c) + (a + b + c)R$$
$$= (a + b + c)(r + R),$$

so $y + z - x = r + R$ as claimed.

We can now outline the proof of the Japanese theorem, using a pentagon inscribed in a circle, as shown in Figure 10.14 (the procedure is the same for any polygon). Since the circle serves as the circumcircle of every triangle in the triangulation, we need only the distances from the circumcenter to the sides (here x, y, z, u, v) and to the diagonals (here s and t).

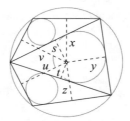

Figure 10.14.

If we let r_1, r_2, and r_3 be the inradii of the three incircles (from, say, smallest to largest), then Carnot's theorem yields $r_1 + R = x + v - s$, $r_2 + R = u + z - t$, and $r_3 + R = y + s + t$. Adding, we have $r_1 + r_2 + r_3 + 3R = x + y + z + u + v$. Thus the sum $r_1 + r_2 + r_3$ is a function of the circumradius and the distances from the center to the five sides, independent of the triangulation. The distance from the circumcenter to a diagonal appears with opposite signs in two occurrences of Carnot's theorem, whereas the distance to each of the sides appears in exactly one instance.

10.3 Tangential and bicentric quadrilaterals

A quadrilateral is *tangential* if it has an incircle, and *bicentric* if it is both cyclic and tangential. If $ABCD$ is a tangential quadrilateral with sides a, b, c, d, (in that order), then $a + c = b + d$, a result that complements the angle relation $A + C = 180° = B + D$ for cyclic quadrilaterals. The proof follows from Figure 10.15, where we see that line segments tangent to a circle from a point outside the circle have the same length, so that if we set $a = x + y$, $b = y + z$, $c = z + t$, and $d = t + x$, then $a + c = b + d$.

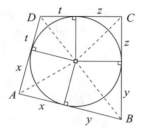

Figure 10.15.

If we let r denote the inradius of the tangential quadrilateral, then its area K is given by $K = r(a + b + c + d)/2 = rs$, where $s = (a + b + c + d)/2$ is the semiperimeter of the quadrilateral. In fact, $K = rs$ holds for every tangential polygon.

There is a method for constructing bicentric quadrilaterals. In a circle, draw two perpendicular chords and construct a quadrilateral $ABCD$ from the tangent lines at the endpoints of the chords, as shown in Figure 10.16.

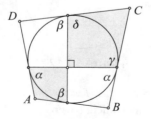

Figure 10.16.

The two semi-inscribed angles marked α are equal since they intercept the same arc of the circle (recall Section 9.2), as are the semi-inscribed angles marked β. Adding the angle measures of the two gray quadrilaterals yields

$A + \alpha + \beta + 180° + \delta + \gamma + C = 720°$. But $\alpha + \gamma = 180°$ and $\beta + \delta = 180°$, thus $A + C = 180°$ so that $ABCD$ is cyclic, and hence is a bicentric quadrilateral.

When the quadrilateral $ABCD$ is bicentric, Brahmagupta's formula for the area K of $ABCD$ (Section 10.1) simplifies to $K = \sqrt{abcd}$. Since the area is also given by $K = rs$, we can express the inradius of a bicentric quadrilateral as a function of the four sides: $r = \sqrt{abcd}/s$.

10.4   Fuss's theorem

In Section 9.4, we encountered Euler's triangle theorem, which provides a relationship among the inradius r, circumradius R, and distance d between the incenter and circumcenter of a triangle: $R^2 - d^2 = 2rR$. This can be written as

$$\frac{1}{R+d} + \frac{1}{R-d} = \frac{1}{r}.$$

For a bicentric quadrilateral (see Figure 10.17) the relationship among r, R, and d is remarkably similar, with the same terms as for a triangle but squared:

$$\frac{1}{(R+d)^2} + \frac{1}{(R-d)^2} = \frac{1}{r^2}. \tag{10.1}$$

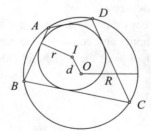

Figure 10.17.

This relationship is known as *Fuss's theorem*, after Nicolaus Fuss (1755–1826), a Swiss mathematician who, upon the recommendation of Daniel Bernoulli, served as Leonhard Euler's secretary at the Academy of St. Petersburg. Our proof of the theorem is from [Salazar, 2006].

In a bicentric quadrilateral $ABCD$ with incenter I and circumscenter O, draw IB, ID, and inradii to sides AB and AD as shown in Figure 10.18a. Since $\angle B + \angle D = 180°$, $\alpha + \beta = 90°$, and the gray shaded triangles are similar, and can be joined to form the right triangle IBD (enlarged) in Figure 10.18b.

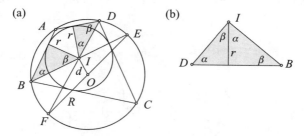

Figure 10.18.

Hence $r|BD| = |IB| \cdot |ID|$ and thus $r^2(|IB|^2 + |ID|^2) = |IB|^2 |ID|^2$, so that

$$\frac{1}{r^2} = \frac{|IB|^2 + |ID|^2}{|IB|^2 |ID|^2} = \frac{1}{|IB|^2} + \frac{1}{|ID|^2}. \tag{10.2}$$

In Figure 10.18a, extend BI to meet the circumcircle at E, and DI to meet the circumcircle at F. Then $\angle COE = 2\angle CBE = 2\alpha$ and $\angle COF = 2\angle CDF = 2\beta$, and hence EOF is a diameter of the circumcircle. Now apply Apollonius's theorem (see Section 6.2) to triangle EIF to conclude

$$|IE|^2 + |IF|^2 = 2(R^2 + d^2). \tag{10.3}$$

The power of the point I with respect to the circumcircle is $|IB| \cdot |IE| = |ID| \cdot |IF| = (R - d)(R + d) = R^2 - d^2$, and hence

$$\frac{1}{|IB|^2} + \frac{1}{|ID|^2} = \frac{|IE|^2}{(R^2 - d^2)^2} + \frac{|IF|^2}{(R^2 - d^2)^2} = \frac{|IE|^2 + |IF|^2}{(R^2 - d^2)^2}. \tag{10.4}$$

Combining (10.2), (10.3), and (10.4) yields (10.1).

10.5   The butterfly theorem

The butterfly theorem is almost 200 years old, having first appeared in 1815. It concerns a surprising property of a *complex* (self-intersecting) quadrilateral inscribed in a circle. The surprise is the unexpected symmetry arising from an almost random construction. Our proof is from [Coxeter and Greitzer, 1967], and is short and direct. For ten additional proofs and more on its history, see [Bankoff, 1987].

The butterfly theorem. *Through the midpoint M of a chord PQ in a circle, any other chords AB and CD are drawn, and chords AD and BC intersect PQ at X and Y, respectively, Then M is the midpoint of XY.* See Figure 10.19a.

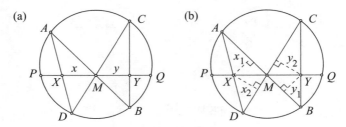

Figure 10.19.

Let $a = |PM| = |MQ|$, $x = |XM|$, $y = |MY|$, and draw line segments x_1 and x_2 from X perpendicular to AB and CD and y_1 and y_2 from Y perpendicular to AB and CD, shown as the dashed lines in Figure 10.19b. Since we have two pairs of equal vertical angles at M as well as equal angles at A and C and at B and D, there are four pairs of similar right triangles, which yield:

$$\frac{x}{y} = \frac{x_1}{y_1}, \quad \frac{x}{y} = \frac{x_2}{y_2}, \quad \frac{x_1}{y_2} = \frac{|AX|}{|CY|}, \quad \text{and} \quad \frac{x_2}{y_1} = \frac{|XD|}{|YB|}.$$

Computing the power of the points X and Y with respect to the circle yields $|AX| \cdot |XD| = |PX| \cdot |XQ|$ and $|CY| \cdot |YB| = |PY| \cdot |YQ|$, whence

$$\frac{x^2}{y^2} = \frac{x_1}{y_1} \cdot \frac{x_2}{y_2} = \frac{x_1}{y_2} \cdot \frac{x_2}{y_1} = \frac{|AX| \cdot |XD|}{|CY| \cdot |YB|} = \frac{|PX| \cdot |XQ|}{|PY| \cdot |YQ|}$$
$$= \frac{(a-x)(a+x)}{(a+y)(a-y)} = \frac{a^2 - x^2}{a^2 - y^2},$$

and $x = y$ as claimed.

10.6 Challenges

10.1. In a square inscribe a circle, and in it inscribe a second square with sides parallel to those of the original square, as shown in Figure 10.20. Show that the area of inner square is one-half that of the original square.

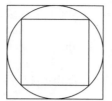

Figure 10.20.

This may have played a role in the construction of some medieval church cloisters, as indicated by this passage form Chapter 12 of Ken Follett's novel *The Pillars of the Earth* [Follett, 1989]:

> "My stepfather, the builder, taught me how to perform certain operations in geometry: how to divide a line exactly in half, how to draw a right angle, and how to draw one square inside another so that the smaller is half the area of the larger."

> "What is the purpose of such skills?" Josef interrupted.

> "Those operations are essential in planning buildings," Jack replied pleasantly, pretending not to notice Josef's tone. "Take a look at this courtyard. The area of the covered arcades around the edges is exactly the same as the open area in the middle. Most small courtyards are built like that, including the cloisters of monasteries. It's because these proportions are most pleasing. If the middle is bigger, it looks like a marketplace, and if it's smaller, it just looks as if there's a hole in the roof. But to get it exactly right, the builder has to be able to draw the open part in the middle so that it's precisely half the area of the whole thing."

10.2. Let a, b, c, d be the sides and p, q the diagonals in a bicentric quadrilateral Q. Prove that $(a + b + c + d)^2 \geq 8pq$.

10.3. In Figure 10.21 triangle AOB is inscribed in a quarter circle with center O. Prove that the diameter of the incircle of $\triangle AOB$ is twice the diameter of the largest circle that can be inscribed in the circular segment adjacent to the hypotenuse of $\triangle AOB$.

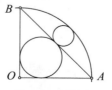

Figure 10.21.

10.4. Prove the following generalization of the butterfly theorem: Let O be any point on a chord PQ of a circle, and draw chords AB and CD through O. Let X and Y be the points of intersection of chords AD and BC with PQ, and set $p = |PO|, q = |OQ|, x = |XO|$, and $y = |OY|$. See Figure 10.22. Prove that $(1/x) - (1/y) = (1/p) - (1/q)$.

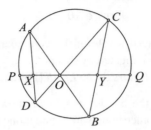

Figure 10.22.

10.5. Let $ABCD$ be a cyclic quadrilateral with perpendicular diagonals that intersect at E. Prove that $|AE|^2 + |BE|^2 + |CE|^2 + |DE|^2 = 4R^2$, where R is the circumradius of $ABCD$.

10.6. Show that a bicentric trapezoid is necessarily isosceles, and that its altitude is the geometric mean of its bases.

CHAPTER **11**

Two Circles

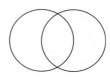

A circle is the reflection of eternity. It has no beginning and it has no end.

Maynard James Keenan

The whole universe is based on rhythms. Everything happens in circles.

John Cowan Hartford

This chapter is devoted to properties of a pair of circles in their many forms: (a) disjoint, (b) tangent, (c) intersecting, or (d) concentric, as illustrated in Figure 11.1

Figure 11.1.

Some of the regions bounded by such circles are sufficiently important to have names in Figure 11.2, such as (a) and (b) *lunes* or *crescents*, (c) a *lens*, (d) the symmetric lens called *vesica piscis*, and (e) the *annulus*.

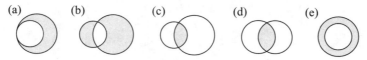

Figure 11.2.

We see objects with these shapes every day. For example the flags of Turkey, Algeria, and at least ten other nations employ the crescent, as does the flag of the state of South Carolina. The logo of a well-known credit card company uses overlapping circles, that of an American television network uses two lunes and a lens, and two annuli form the logo of a large American discount retail corporation, as illustrated in Figure 11.3. Even some food items have these shapes, such as croissants, crescent dinner rolls, and bagels.

Figure 11.3.

In this chapter we discuss a variety of geometric properties related to the various two-circle icons presented above.

11.1 The eyeball theorem

A very pleasing theorem about a pair of circles is

The eyeball theorem. *Suppose two nonintersecting circles are centered at P and Q, and lines through P tangent to the circle centered at Q cut the circle centered at P at points A and B. Similarly, lines through Q tangent to the circle centered at P cut the circle centered at Q at points C and D. Then* $|AB| = |CD|$.

See Figure 11.4.

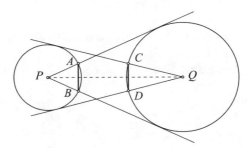

Figure 11.4.

There are many proofs. Ours is from [Konhauser et al., 1996]. In Figure 11.5 we have the upper half of Figure 11.4, where we have set $x = |AB|/2$, $y = |CD|/2$, $d = |PQ|$, and we let r and s denote the radii of the two circles as shown. To show $|AB| = |CD|$ we need only show $x = y$.

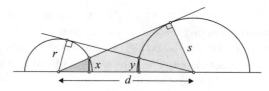

Figure 11.5.

Since the small gray shaded right triangle is similar to the large shaded right triangle, $x/r = s/d$ and hence $x = rs/d$. Similarly $y = rs/d$, thus $x = y$.

11.2 Generating the conics with circles

In this section we show that the ellipse, the parabola, and the hyperbola can be generated using circles. Using the ideas of a Swiss secondary school teacher named Jean-Louis Nicolet, Caleb Gattegno created *Animated Geometry* in 1949, consisting of 22 short (2–5 minutes) silent films illustrating mathematical concepts. The following characterizations of the conics are from the final film, entitled *Common generation of conics* [Gattegno, 1967].

The ellipse. *Let C be a circle and P a point inside C. Then the locus of the centers of circles passing through P and tangent to C is an ellipse with foci P and the center of C.*

See Figure 11.6a, where we see the variable circle (in gray) in several positions inside C.

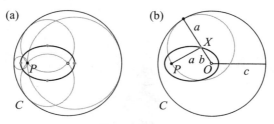

Figure 11.6.

The justification is best accomplished starting with an ellipse with foci O and P, as in Figure 11.6b. If X is a point on the ellipse, then the distances a and b from P and O, respectively, to X satisfy $a + b = c$ for some constant c. With O as center draw the circle C of radius c, and extend OX to meet C.

Thus X is equidistant from P and C, i.e., it is the center of the circle with center X and radius a, passing through P and tangent to C.

A quote from Victor Hugo (1802–1885) is apropos here: "Mankind is not a circle with a single center but an ellipse with two focal points of which facts are one and ideas the other."

The parabola.   *Let C be a circle and L a line not intersecting C. Then the locus of the centers of circles tangent to both C and L is a parabola.*

See Figure 11.7.

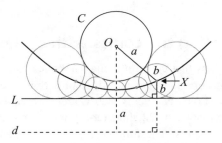

Figure 11.7.

The justification is immediate in the figure, since the distance between the center X of the variable circle to the center O of C is the same as the distance between X and a line d (the directrix) parallel to L.

The hyperbola.   *Let C be a circle and P a point outside C. Then the locus of the centers of circles passing through P and tangent to C is (one branch of) a hyperbola with foci P and the center O of C.*

See Figure 11.8.

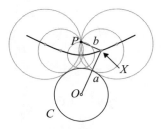

Figure 11.8.

The distance a from the center O of C to the point X on the curve minus the distance b from X to P is the radius of C. Hence the curve is one branch of a hyperbola with foci O and P. The other branch is obtained by using P as the center of C.

Caleb Gattegno and Jean-Louis Nicolet

Caleb Gattegno (1911–1988) was one of the most influential and prolific mathematic educators of the twentieth century. Born in Egypt to Spanish parents, he earned doctorates in mathematics and psychology in Switzerland and France, respectively. Gattegno believed that success in mathematics was everyone's birthright, and emphasized the role of psychology and manipulative devices. He invented the geoboard and was instrumental in popularizing Cuisenaire rods in mathematics education. He founded or co-founded several European mathematics organizations and journals, and wrote over one hundred books.

The first person to suggest the use of films with animated cartoons to teach elementary geometry was Jean-Louis Nicolet, a Swiss mathematics teacher. Gattegno advocated the use of Nicolet's films, and later, in the 1940s made his own to provide a dynamic approach to geometry in the classroom [Powell, 2007].

11.3 Common chords

When two circles intersect they share a common chord, which is a key element in several surprising results.

For example, consider a circle C_1 that passes through the center O of a second circle C_2. Then the length of the common chord PQ is the same as the length of the portion of the tangent line to C_1 at P (or Q) that lies inside C_2. See Figure 11.9 [Eddy, 1992].

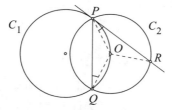

Figure 11.9.

Since $\angle OPR = \angle OQP$ (see Section 9.2), isosceles triangles OPR and OQP are congruent and thus $|PR| = |PQ|$.

Suppose circles C_1 and C_2 intersect with a common chord PQ, as shown in Figure 11.10a. Project point A on the arc of C_1 outside C_2 through P and Q to determine a chord BC of C_2. The surprise is that the length of BC is the same for any position of A on its arc of C_1 [Honsberger, 1978].

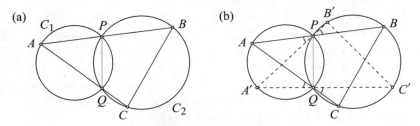

Figure 11.10.

To see this, choose another point A' on the arc of C_1 outside C_2 and project it through P and Q to determine the chord $B'C'$. Since the four angles marked $\angle$ in Figure 11.10b are equal, arcs $B'B$ and $C'C$ have the same length and thus arcs $B'C'$ and BC are equal. Hence $|BC| = |B'C'|$.

Next consider the intersecting circles C_1 and C_2 with common chord PQ and all possible line segments through P with endpoints on the two circles. The longest such segment is the segment AB perpendicular to PQ at P. See Figure 11.11.

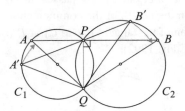

Figure 11.11.

To justify this, consider another segment $A'B'$ through P, draw $A'Q$ and $B'Q$, and since $\angle B'A'Q = \angle BAQ$ and $\angle A'B'Q = \angle ABQ$, triangle $A'B'Q$ is similar to triangle ABQ. Since BQ is a diameter of C_2, $|B'Q| \le |BQ|$. Thus, because of the similarity of the two triangles, $|A'B'| \le |AB|$.

11.4   Vesica piscis

When two circles intersect, the convex region bounded by two circular arcs
is called a *lens*. The lens will have an axis of symmetry when the two circles
have the same radius. If in addition each circle passes through the center of
the other, as in Figure 11.2d, the lens is called a *vesica piscis*, Latin for "fish
bladder." In Italian it is called a *mandorla* ("almond").

The vesica piscis can be seen in the first geometric construction in Euclid's
Elements—Proposition 1 in Book I—the construction of an equilateral triangle. It also arises in bisecting a given line segment, and drawing a line
perpendicular to a given line. See Figure 11.12.

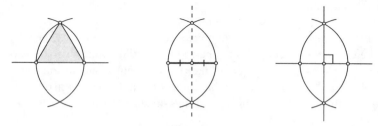

Figure 11.12.

The area K of the vesica piscis constructed from circles of radius r is easily computed as the sum of the areas of two circular sectors each subtended
by an angle of $2\pi/3$ minus the sum of the areas of two equilateral triangles
with base r to yield $K = (2\pi/3 - \sqrt{3}/2)r^2$.

The vesica can also be used to trisect a line segment. In the construction
in Figure 11.13 [Coble, 1994], let AB be the segment to be trisected and
draw the vesica with circles centered at A and B as shown. Then draw the
segments $CD, DB, AE,$ and CF. Since AB and CF are medians of triangle
BCD, we have $|AX| = (1/3)|AB|$.

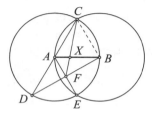

Figure 11.13.

The vesica piscis has a long history as a mystical and religious symbol. It was used in medieval, Romanesque, and Byzantine art as an enclosure or frame for paintings of important personages or sacred events.

In architecture, one form of a gothic arch is the *equilateral arch*, the upper one-half of a vesica piscis. See Challenge 11.6.

Vesica piscis and the ellipse

Beginning in the sixteenth century, churches and cathedrals began to be built in Italy and Spain incorporating the ellipse into their design. But often ellipses were approximated by figures using four or more circular arcs. Sebastiano Serlio (1475–1554), in his classic work *Tutte l'Opere d'Architettura*, describes four such constructions, one of which is based on the vesica piscis. Serlio recommended it for its simplicity, beauty, and ease of construction. To the pair of interesting circles we add two circular arcs, using the vertices of the vesica as centers and the diameter of the circles as radii, as shown in Figure 11.14a. For comparison, Figure 11.14b shows a true ellipse with the same dimensions. For further details, see [Rosin, 2001].

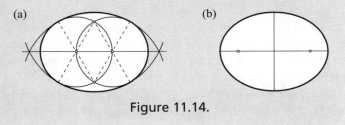

Figure 11.14.

11.5 The vesica piscis and the golden ratio

If we enclose the vesica piscis in a pair of intersecting circles whose centers coincide with the centers of the vesica piscis circles, as shown in Figure 11.15, we have another appearance of the golden ratio ϕ: $|CX|/|CD| = \phi$ [Hofstetter, 2002].

After drawing a pair of circles with radius $|AB|$ and centers A and B, draw two more circles with radius $|AF| = |BE|$, also centered at A and B. Let O be the midpoint of AB, D and X intersection points of the circles as shown in the figure, then O, D, and X are collinear. If we let $|OA| = |OB| = 1$,

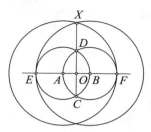

Figure 11.15.

then $|AC| = |AB| = 2$ and $|AX| = |AF| = 4$, so $|CO| = \sqrt{3}$ and $|OX| = \sqrt{15}$. Hence we have

$$\frac{|CX|}{|CD|} = \frac{|CO| + |OX|}{|CD|} = \frac{\sqrt{3} + \sqrt{15}}{2\sqrt{3}} = \frac{1 + \sqrt{5}}{2} = \phi.$$

11.6 Lunes

A *lune* (from the French word for moon) is a concave region in the plane bounded by two circular arcs. See Figure 11.2b for an illustration of two lunes (in gray) and a lens (in white). Lunes are also referred to as crescents, although some authors reserve the word crescent for a lune whose smaller circle contains the center of the larger circle, such as the one in Figure 11.2a.

Hippocrates of Chios (c. 470–410 BCE) is believed to be the first person to square lunes. His success gave hope for squaring the circle, one of the three great geometric problems of antiquity.

Although Hippocrates lived before the time of Euclid, he knew the generalization of the Pythagorean theorem given in Proposition 31 of Book VI: *In right-angled triangles the figure on the side opposite the right angle equals the sum of the similar and similarly described figures on the sides containing the right angle.* Hippocrates used semicircles on the sides of a triangle to prove the following theorem: *If a square is inscribed in the circle and four semicircles constructed on its sides, then the area of the four lunes equals the area of the square.* See Figure 11.16.

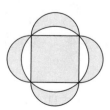

Figure 11.16.

For a visual proof, see Figure 11.17.

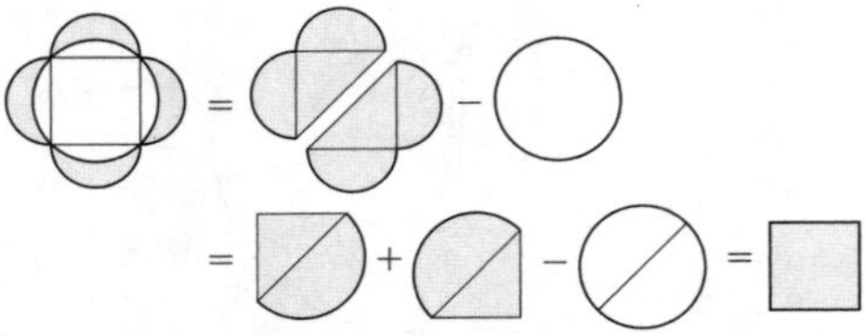

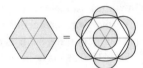

Figure 11.17.

Dividing Figure 11.16 into two congruent halves along a diagonal of the square shows that the area of two lunes on the legs of an isosceles right triangle equals the area of the triangle. Hippocrates proved that the same is true for any right triangle, as in Challenge 4.7.

Hippocrates also established the following result relating the areas of a hexagon, six lunes, and a circle: *If a regular hexagon is inscribed in a circle and six semicircles constructed on its sides, then the area of the hexagon equals the area of the six lunes plus the area of a circle whose diameter is equal to one of the sides of the hexagon.* See Figure 11.18.

Figure 11.18.

In the visual proof in Figure 11.19 [Nelsen, 2002c], we use the fact (known to Hippocrates) that the area of a circle is proportional to the square of its radius, so that the four small gray circles in the second line have the same combined area as the large white circle.

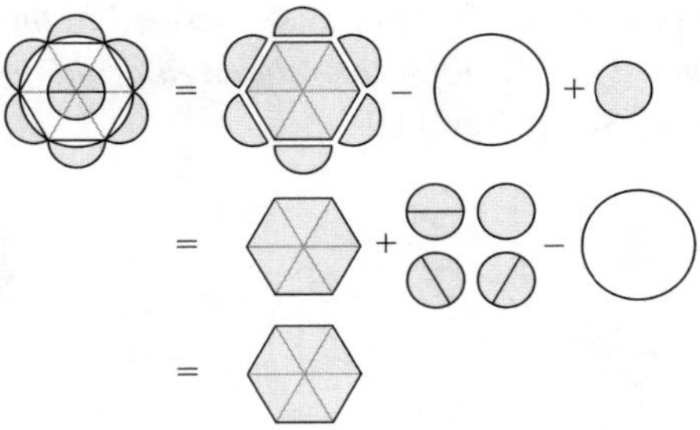

Figure 11.19.

11.7   The crescent puzzle

One form of a crescent is the region between two circles, one interior and tangent to the other. It is an artistic representation of the moon in the first or third quarter. Although the crescent is often associated with Islam, it is an ancient symbol, predating Islam by many centuries.

Crescents figure prominently in recreational mathematics problems. One such is "the crescent puzzle," problem 191 in Henry Dudeney's 1917 classic, *Amusements in Mathematics*. In it we are given the width of the crescent at two places, as shown in Figure 11.20a, and asked to find the diameters of the circles.

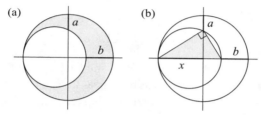

Figure 11.20.

If we let x be the radius of the larger circle, then the two diameters are $2x$ and $2x - b$. The right triangle altitude theorem in Section 4.2 tells us that $x - a$ is the geometric mean of x and $x - b$. Thus $x - a = \sqrt{x(x - b)}$, and consequently $x = a^2/(2a - b)$. Hence the two diameters are $2a^2/(2a - b)$ and $[2a^2/(2a - b)] - b$.

11.8   Mrs. Miniver's problem

Mrs. Miniver is a fictional character created by the British author Joyce Maxtone Graham (1901–1953), who wrote columns under the pen name Jan Struther for *The Times* of London between 1937 and 1939. In a column entitled "A Country House Visit," she describes an aspect of real-life relationships in mathematical terms [Struther, 1990]:

> "She saw every relationship as a pair of intersecting circles. The more they intersected, it would seem at first glance, the better the relationship; but this is not so. Beyond a certain point the law of diminishing returns sets in, and there aren't enough private resources left on either side to enrich the life that is shared. Probably perfection is reached

when the area of the two outer crescents, added together, is exactly equal to that of the leaf-shaped piece in the middle. On paper there must be some neat mathematical formula for arriving at this: in life, none."

What is the solution to Mrs. Miniver's problem, remembering that two circles are rarely equal? See Figure 11.21.

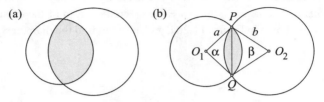

Figure 11.21.

Let the circles have radii a and b, $a \leq b$, and let L denote the area of the leaf (i.e., the lens). If the two crescents have areas C_1 and C_2, then $C_1 + C_2 + 2L = \pi(a^2 + b^2)$. Perfection in the relationship requires $L = C_1 + C_2$ and hence $3L = \pi(a^2 + b^2)$. Since the maximum value of L is πa^2, we have $a \leq b \leq a\sqrt{2}$.

A *circular segment,* the region between a chord and an arc subtended by the chord, has area $R^2(\theta - \sin\theta)/2$ where R is the radius of the circle and θ is the angle in radians at the center subtended by the arc. The leaf consists of two circular segments, subtended by angles α and β as shown in Figure 11.10b, and hence

$$L = \frac{a^2}{2}(\alpha - \sin\alpha) + \frac{b^2}{2}(\beta - \sin\beta).$$

But α and β are related by $|PQ| = 2a\sin(\alpha/2) = 2b\sin(\beta/2)$, and so if we let $r = b/a$ be the ratio of the two radii (with $1 \leq r \leq \sqrt{2}$) and recall that $L = \pi(a^2 + b^2)/3$, we have, after some algebra,

$$\frac{\pi}{3}(1+r^2) = \frac{r^2}{2}(\beta - \sin\beta) + \arcsin\left(r\sin\frac{\beta}{2}\right) - \frac{1}{2}\sin\left(2\arcsin\left(r\sin\frac{\beta}{2}\right)\right).$$

Given r, the equation can be solved numerically for β. If $r = 1$ (i.e., $a = b$ and $\alpha = \beta$) we have $\beta - \sin\beta = 2\pi/3$, and β is approximately 2.6053256746 radians, or 149° 16′ 27″. The distance between the centers of the two circles is approximately $0.529864a$.

11.9 Concentric circles

An *annulus* is the region between two circles with the same center but different radii. The area of an annulus is the same as the area of a circle whose diameter is a chord of the outer circle tangent to the inner circle, as illustrated in Figure 11.22.

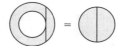

Figure 11.22.

If we let the radii of the outer and inner circles be a and b, respectively, $a > b$, then the area of the annulus is $\pi(a^2 - b^2)$. The length of the chord is $2\sqrt{a^2 - b^2}$, and hence a circle with this diameter has the same area as the annulus.

The bull's eye illusion
Which region in Figure 11.23 appears to have the greater area—the inner white disk or the outer white annulus?

Figure 11.23.

At first glance the disk in the center may appear larger in area than the annulus, but the two have the same area. But if we let the radii of the circles be 1, 2, 3, 4, and 5, then the area of the annulus is $(5^2 - 4^2)\pi = 3^2\pi$, the same as that of the inner disk [Wells, 1991].

Suppose an annulus of outer radius a and inner radius b has the same area as an ellipse with semi-major and semi-minor axes of lengths a and b, respectively, as illustrated in Figure 11.24. What can we say about the ratio of a to b?

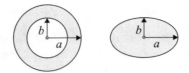

Figure 11.24.

Since the area of the annulus is $\pi(a^2 - b^2)$ and the area of the ellipse is πab, the areas are the same if and only if $a^2 - ab - b^2 = 0$, or equivalently, $(a/b)^2 - (a/b) - 1 = 0$. Since $a/b > 0$, a/b must be the golden ratio $\phi = (1 + \sqrt{5})/2$. Such an ellipse is sometimes called a *golden ellipse*, as it can be inscribed in a golden rectangle, in this case a $2a \times 2b$ rectangle [Rawlins, 1995].

Bertrand's paradox

Joseph Louis François Bertrand (1822-1900) introduced the following problem in his book *Calcul des Probabilités* in 1889: Given two concentric circles with radii r and $2r$, what is the probability that a chord drawn at random in the larger circle will intersect the smaller circle (as in Figure 11.25a)?

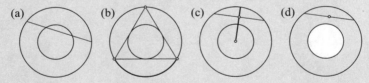

Figure 11.25.

The answer depends on how we select a chord "at random." After choosing one end of the chord, the other end must be in the middle third of the circumference as in Figure 11.25b, so the probability is $1/3$. If we focus on the midpoint of the chord, its distance from the center must be less than half way to the outer circle, so the probability is $1/2$, as in Figure 11.25c. Or the center must lie inside the smaller circle, so the probability is $1/4$, the ratio of the areas of the two circles, as seen in Figure 11.25d.

11.10 Challenges

11.1. Two circles with centers P and Q are tangent externally at A as shown in Figure 11.26. If the line segment BC is tangent to both circles, show that $\angle BAC = 90°$.

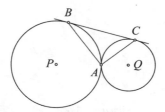

Figure 11.26.

11.2. Two unit circles are tangent externally, as shown in Figure 11.27. From a point P on one circle rays PQ and PR are drawn, intersecting both circles. If x, y, and z denote the arc lengths of the two circles lying between the rays, prove that $x + y = z$.

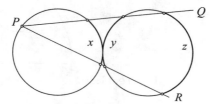

Figure 11.27.

11.3. From point A exterior to a circle draw two lines tangent to the circle at B and C, as shown in Figure 11.28. Show that the incenter of triangle ABC lies on the given circle.

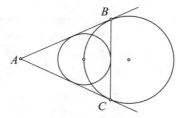

Figure 11.28.

11.4. Figure 11.29 shows a circular lamina of radius 1 from which is removed a circular lamina of radius x. If the center of gravity of the remaining crescent is at the edge of the removed lamina (the black dot in the figure), show that $x = 1/\phi \approx 0.618$, where $\phi = (1 + \sqrt{5})/2$ is the golden ratio. (In [Glaister, 1996] this crescent is called a *golden earring*.)

Figure 11.29.

11.5. Line segments whose lengths are the square roots of the first five positive integers can be found in the vesica piscis diagram constructed from unit circles. Can you locate them?

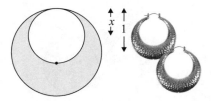

Figure 11.30.

11.6. The gothic equilateral arch is based on the upper half of the vesica piscis, and is often embellished with smaller arches and rose windows, as illustrated on the 20 euro banknote in Figure 11.30. In the design suppose we wish to locate a circular rose window tangent to each of the arches. Locate its center and radius. [Hint: each of the smaller arches is similar to the larger one.)

11.7. Suppose P is a point inside two concentric circles, different from the common center. A ray from P intersects the inner circle at Q and the outer circle at R, as shown in Figure 11.31. For which direction from P is the length of QR a maximum?

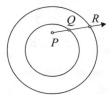

Figure 11.31.

11.8. Given a circle C with center O and radius r, let P be an exterior point. If a ray is drawn from P intersecting the circle at A and B, what is the locus of the midpoint M of the chord AB?

11.9. Consider two externally tangent circles with centers P and Q, radii r and r', respectively, and common tangents meeting at V, as shown in Figure 11.32. Show that the radius R of the circle tangent to the two lines with center at the point of tangency of the two circles is the harmonic mean of r and r'.

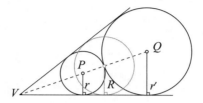

Figure 11.32.

11.10. Consider two externally tangent circles with centers P and Q, radii r_1 and r_2, and a common tangent line, as shown in Figure 11.33. Show that the distance $|AB|$ between the points of tangency is twice the geometric mean of r_1 and r_2.

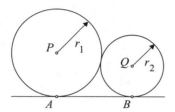

Figure 11.33.

11.11. Suppose a third circle with center R and radius r_3 is tangent to the two circles and the line in Challenge 11.10, as shown in gray in Figure 11.34. Prove that

$$\frac{1}{\sqrt{r_3}} = \frac{1}{\sqrt{r_1}} + \frac{1}{\sqrt{r_2}}.$$

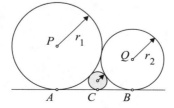

Figure 11.34.

CHAPTER 12

Venn Diagrams

> *Our task must be to free ourselves from this prison by widen-*
> *ing our circles of compassion to embrace all living creatures*
> *and the whole of nature in its beauty.*
>
> Albert Einstein

John Venn (1834–1923) introduced the diagrams that bear his name in his paper entitled *On the Diagrammatic and Mechanical Representation of Propositions and Reasonings* in 1880 [Venn, 1880]. The diagrams, which Venn called "Eulerian circles," were used by Venn to represent sets and the relationships among them. They became a common part of the new math movement of the 1960s based on set theory. Venn diagrams usually consist of intersecting circles (as in our icon), although other shapes may be used. Similar diagrams can be found in the work of Ramon Llull (1232–1315), Gottfried Wilhelm Leibniz (1646–1716), and Leonhard Euler (1708–1783).

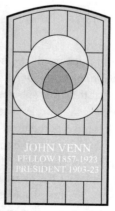

John Venn and stained glass in Caius Hall, Cambridge University

149

Images of three or more intersecting circles may be found in gothic windows, paintings, graphical designs, and in the logo of the Olympic games.

We will look at the role played by Venn diagrams, i.e., intersecting circles, in geometry rather than their traditional application to logic and set theory. After considering some results about three circles we proceed to look at triangles with intersecting circles. We conclude with some remarks about figures related to Venn diagrams such as Reuleaux triangles and Borromean rings.

12.1   Three-circle theorems

We begin by considering some theorems about three intersecting circles. The results primarily concern properties of the points of intersection.

Suppose we have three circles in the plane, each intersecting the other two twice, but with no point common to all three, such as in the Venn diagram or, more generally, as in Figure 12.1. If we draw the common chords for each pair of the circles, they meet in a point [Bogomolny, 2010].

Figure 12.1.

To prove this nice result, we need a preliminary three-dimensional result: *three spheres intersecting one another have at most two points in common.* We assume that the three spheres are in "general position," i.e., each pair of spheres intersect in a circle. The proof is easy—two spheres intersect in a circle, which intersects the third sphere in two points. Consider Figure 12.1 embedded in space, where the three circles are the equators of three spheres cut by the plane of the page. Then the chords are the projections onto the page of the circles of the pairwise intersections of the spheres. By our preliminary result, the three spheres meet in two points, both of which project onto the point of intersection of the three chords.

The common chord for a pair of intersecting circles is called the *radial axis* for the circles, and the point of intersection of the three radical axes for three circles considered pairwise (as above) is called the *radical center* of the circles. The radical axes and center help us prove the following theorem of Hiroshi Haruki.

Haruki's theorem. *Suppose each of three circles intersects each of the others in two points. If we label the line segments as shown in Figure 12.2, then*

$$\frac{a}{b} \cdot \frac{c}{d} \cdot \frac{e}{f} = 1.$$

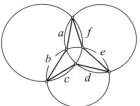

Figure 12.2.

To prove the theorem [Honsberger, 1995], draw the three radical axes, label the points as shown in Figure 12.3a, and let x, y, and z denote the lengths of segments PD, PE, and PF, respectively.

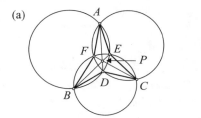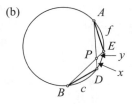

Figure 12.3.

Consider the portion of Figure 12.3a lying in the circle through A and B, as shown in Figure 12.3b. Triangles AEP and BDP are similar and thus $f/y = c/x$. Similarly, chords in the other two circles yield $b/z = e/y$ and $d/x = a/z$. The product of the proportions yields $fbd/yzx = cea/xyz$ and hence $bdf = ace$, which proves the theorem.

In 1916 R. A. Johnson discovered the following result [Johnson, 1916], which has been described as one of the few recent "really pretty theorems at the most elementary level of geometry" [Honsberger, 1976], one we call

Johnson's theorem. *If three circles with the same radius are drawn through a point, then the other three points of intersection determine a fourth circle with the same radius.*

See Figure 12.4a.

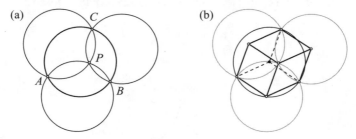

Figure 12.4.

The elegant proof is based on looking at the configuration from a three dimensional point of view. Label the points as shown in Figure 12.4a, and let r denote the common radius of the circles. The points A, B, C of intersection and the centers of the circles form a hexagon divided into three rhombi, as illustrated in Figure 12.4b. The nine dark line segments each have length r, and drawing the three dashed line segments of length r produces a plane projection of a cube. Hence A, B, and C lie at a distance r from another point, and hence the fourth circle also has radius r.

Our next three-circle theorem is attributed to Gaspard Monge (1746–1818).

Monge's theorem. *Consider three circles C_1, C_2, C_3, such that C_1 and C_2 intersect at Q and R and C_2 and C_3 intersect at S and T. If ray RQ and ray TS meet at P, as shown in Figure 12.5a, then the tangents from P to the three circles all have the same length.*

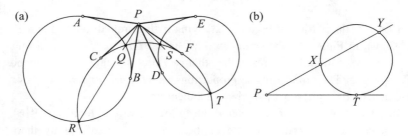

Figure 12.5.

We will show $|PA| = |PB| = |PC| = |PD| = |PE| = |PF|$. Our tool is the power of a point with respect to a circle from Section 9.3, which states that if a tangent PT and PXY are a tangent and a secant to a circle as shown in Figure 12.5b, then $|PT|^2 = |PX||PY|$. Thus

$$|PA|^2 = |PB|^2 = |PQ||PR| = |PC|^2 = |PF|^2$$
$$= |PS||PT| = |PD|^2 = |PE|^2$$

and the result follows.

In Section 11.4 we used two intersecting circles (the vesica piscis diagram) to trisect a line segment. Three identical circles, each passing through the centers of the other two, can also be used [Styer, 2001]. See Figure 12.6.

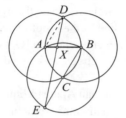

Figure 12.6.

To trisect AB, draw three circles with a common radius $|AB|$ and centers A, B, C as shown. Draw segment BE through C and then DE intersecting AB at X. Since AD is parallel to BE, triangles ADX and BEX are similar, with $|BE| = 2|AD|$. Hence $|BX| = 2|AX|$ and $|AX| = (1/3)|AB|$ as desired.

12.2 Triangles and intersecting circles

We have seen a variety of theorems about triangles and circles or semicircles in Chapters 4 and 7. We now consider some theorems concerning triangles and three intersecting circles.

In Figure 12.7a we see a right triangle with circles drawn whose centers and diameters are the midpoints and lengths of the three sides. If T denotes the area of the triangle, A the area of the arbelos-like region below the hypotenuse, and $B = B_1 + B_2$ the area of the lens-shaped region inside the triangle, then $T = A - B$. See Figure 12.7b.

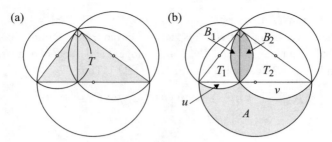

Figure 12.7.

Let T_1 denote the area of the right triangle to the left of the altitude to the hypotenuse in the large triangle, and T_2 the area of the right triangle to the right of the altitude, so that $T = T_1 + T_2$. Then $B_2 + T_1 + u$ and $B_1 + T_2 + v$ are the areas of semicircles drawn on the legs of the triangle, so they sum to the area $A + u + v$ of the semicircle drawn on the hypotenuse (by Proposition IV.31 in Euclid's *Elements,* cited in the proof of Archimedes' Proposition 4 in Chapter 4). Hence $B + T = A$ as claimed [Gutierrez, 2009].

For a similar result about five circles associated with a right triangle, see Challenge 12.1.

In 1838 the French mathematician August Miquel published the following theorem, which now bears his name. It is sometimes referred to as the pivot theorem.

Miquel's theorem. *In triangle ABC, let P, Q, and R be points on sides AB, BC, and CA, as illustrated in Figure 12.8a. Then the circumcircles of triangles APR, BPQ, and CQR all pass through a common point M (sometimes called the Miquel point of the three circles).*

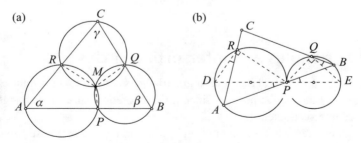

Figure 12.8.

Consider the circles that pass through the vertices A and B intersecting at P. If they also intersect at a second point M, (as in Figure 12.8a), we claim that M also lies on the circle through C, R, and Q. Since $\angle PMR = 180° - \alpha$ and $\angle PMQ = 180° - \beta$, we have

$$\angle QMR = 360° - (\angle PMR + \angle PMQ) = \alpha + \beta = 180° - \gamma.$$

Thus $CRMQ$ is a cyclic quadrilateral so that M lies on the circle through C, R, and Q as claimed.

If the two circles through A and B are tangent at P, we claim that P lies on the circle through C, R, and Q. See Figure 12.8b, and draw the diameters DPE of the two circles. Assume DE intersects AC (the case when DE intersects BC is similar). Let x denote the measure of each of the four marked angles $\angle ARD$, $\angle APD$, $\angle BPE$, and $\angle BQE$. Since triangles DRP and EQP are right triangles, we have

$$\angle CRP + \angle CQP = (90° + x) + (90° - x) = 180°.$$

Thus $CRPQ$ is a cyclic quadrilateral so that P lies on the circle through C, R, and Q as claimed.

12.3   Reuleaux polygons

A *Reuleaux triangle* is the central portion of the Venn diagram common to all three circles. It can be easily drawn starting with an equilateral triangle and adding three circular arcs whose centers are the vertices and whose radii are the sides of the triangle. Reuleaux triangles have been employed in Gothic architecture from the Middle Ages to the present. The window in Figure 12.9a is from the Església Mare de Déu de Montsió in Barcelona, Spain, and the window in Figure 12.9b is from the Scots Church in Adelaide, Australia.

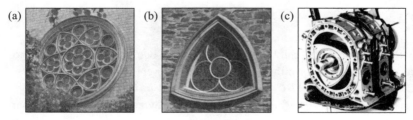

(a)   (b)   (c)

Figure 12.9.

The German engineer Franz Reuleaux (1829–1905) studied this "circular triangle" in relation with rotating mechanisms. The rotor in the Wankel rotary engine is in the shape of a Reuleaux triangle. See Figure 12.9c.

The *width* of a closed convex curve is defined as the maximum distance between two opposite parallel lines touching its boundary. The Reuleaux triangle is an example of a *curve of constant width*, that is, its width is the same in every direction (like the circle).

Reuleaux polygons are constructed analogously to the triangles, replacing the equilateral triangle by a regular polygon with an odd number of sides. Every Reuleaux polygon is a curve of constant width. *Barbier's theorem* states that every curve of constant width w has the same perimeter πw (this is immediate in the case of a Reuleaux polygon with an odd number of sides), and the *Blaschke-Lebesgue theorem* says that among all curves of constant width, the one with the smallest area is the Reuleaux triangle.

Reuleaux polygons and coin design

Reuleaux polygons are sometimes employed by mints to produce non-circular coins that function in coin-operated machines, and enable people with limited sight to distinguish coins by touch. In Figure 12.10 we see a seven-sided fifty pence coin of the United Kingdom, a nine-sided five euro coin from Austria, and an eleven-sided dollar coin from Canada.

Figure 12.10.

Reuleaux triangles can be generalized to non-equilateral triangles and to curves that do not have the angular corners present in Reuleaux triangles. Consider a triangle ABC with sides a, b, and c. Choose k so that $k > \max\{a + b, b + c, c + a\}$ and extend the sides beyond the vertices as shown in Figure 12.11 (although we have illustrated the procedure for the 3-4-5 right triangle with $k = 10$, the procedure works for an arbitrary triangle). Draw circular arcs using each vertex as a center. For example, with vertex B as the center draw arcs with radii $k - c - a$ and $a + (k - a - b) = k - b$.

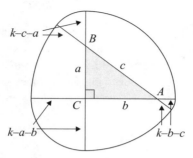

Figure 12.11.

The resulting curve has constant width $2k - a - b - c$ and is smooth, in the sense that a tangent line exists at each point on the curve.

Borromean rings

The *Borromean rings*, a Venn diagram-like object studied in topology and knot theory, consist of three interlocking rings as illustrated in Figure 12.12a. The rings have the property that taken together they are inseparable, but if one ring is removed the other two are not linked. The name comes from the House of Borromeo in Milan, whose family crest has incorporated the rings since the 15th century. See Figure 12.12b. In the United States they are sometimes called the *Ballantine rings*, from their use in the logo of Ballantine beer and ale. They also appear on the logo of the Ricordi music publishing company of Milan, whose bicentennial in 2008 was marked with the postage stamp in Figure 12.12c.

Figure 12.12.

The edges of the three mutually perpendicular golden rectangles found in the regular icosahedron illustrated in Figure 4.19 (and reproduced in Figure 12.13a) are linked like the Borromean rings, and lead to the noncircular Borromean links of the logo of the International Mathematical Union in Figure 12.13b. But do Borromean *circles* exist in R^3? The surprising answer is no, no matter what their size or orientation in R^3. See [Lindström and Zetterström, 1991] for a proof.

(a)   (b)

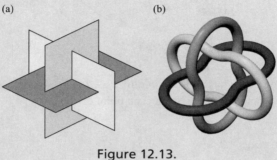

Figure 12.13.

12.4 Challenges

12.1. For a right triangle, draw the circles whose centers and diameters are the midpoints and lengths of the following five line segments: the two legs, the altitude to the hypotenuse, and the two segments of the hypotenuse determined by the foot of that altitude, as shown in Figure 12.14. Prove that the sum $A + B + C + D$ of the areas of the four shaded lunes is equal to the area T of the triangle. (Hint: See Challenge 4.7.)

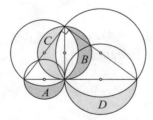

Figure 12.14.

12.2. Given a triangle ABC and a point P on its circumcircle, draw the perpendiculars PQ, PR, and PS to the sides AB, BC, and AC

(extended if necessary) respectively. Show that Q, R, and S lie on the same line. See Figure 12.15.

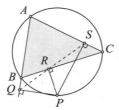

Figure 12.15.

12.3. Show that in the plane the maximum number of regions determined by n circles is $n^2 - n + 2$.

12.4. Let A and B be the points of intersection of two circles. How many lines through A yield chords of equal length in the two circles?

12.5. Four quarter circles are inscribed in a square with side 1 as shown in Figure 12.16. Find the area of the region common to all four. (No calculus, analytic geometry, or trigonometry is required.)

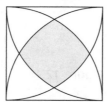

Figure 12.16.

12.6. In "The Bun Puzzle" in his *Amusements in Mathematics*, Henry Ernest Dudeney [Dudeney, 1917] asks us to solve the following puzzle (see Figure 12.17): "The three circles represent three buns, and it is simply required to show how these may be equally divided among four boys. The buns must be regarded as of equal thickness throughout and of equal thickness to each other. Of course, they must be cut into as few pieces as possible. To simplify it I will state the rather surprising fact that only five pieces are necessary, from which it will be seen that one boy gets his share in two pieces and the other three receive theirs in a single piece." [Hint: the ratio of the diameters of the buns is 3-4-5.]

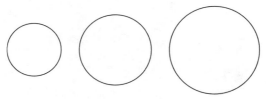

Figure 12.17.

12.7. Given three identical circles each tangent to the other two, find the area of the shaded region enclosed by the arcs joining the points of tangency, as shown in Figure 12.18.

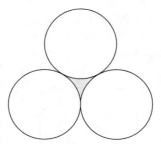

Figure 12.18.

12.8. Is it possible to assign the numbers 1, 2, 3, 4, 5, 6, 7 to the seven regions of the Venn diagram so that the sum of the numbers in each circle is the same?

12.9. Let P be a point in the plane of an equilateral triangle ABC such that each of the triangles PAB, PBC, and PCA is isosceles. How many positions are there for the point P?

12.10. Show that the sides of the Napoleon triangle formed by the centroids of the equilateral triangles on the sides of $\triangle ABC$ are the perpendicular bisectors of the segments AF, BF, and CF joining the Fermat point F to the vertices of $\triangle ABC$. See Figure 12.19. (Hint: Consider the circumcircles of the three shaded equilateral triangles.)

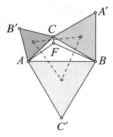

Figure 12.19.

CHAPTER 13

Overlapping Figures

> *A square is neither line nor circle; it is timeless. Points don't chase around a square. Firm, steady, it sits there and knows its place. A circle won't be squared.*

Unknown

In many of the preceding chapters we have used icons that consist of a geometric figure partitioned into several other figures. In this chapter we extend that idea to figures that overlap. For example, we interpret the icon above as consisting of two overlapping squares within a larger square rather than a square partitioned into three smaller squares and two L-shaped regions.

This simple idea has remarkable consequences. In this chapter we first introduce the little-known carpets theorem and present two applications—a proof of the irrationality of $\sqrt{2}$ and a characterization of Pythagorean triples. Overlapping figures also provide a natural way to illustrate a variety of inequalities.

(a)   (b)

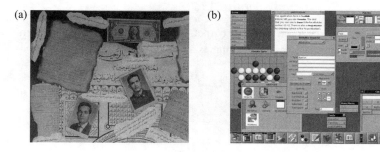

Figure 13.1.

Overlapping figures appear in many works of art from classical paintings to abstract works such those of Joan Miró and Paul Klee. A *collage* is an artistic composition often constructed from overlapping prints, photographs,

163

and cutouts, such as the one in Figure 13.1a. An entire branch of mathematics is devoted to overlapping figures—knot theory. We also see overlapping figures every day on computer screens with many windows open simultaneously, as in Figure 13.1b.

13.1   The carpets theorem

A simple but powerful result for solving problems concerning overlapping figures is *the carpets theorem*. Suppose we have a room with two carpets that completely cover the floor, as illustrated in Figure 13.2a. If we move one of them, as in Figure 13.2b, then the area of the overlap (in dark gray) must equal the uncovered area (in white). This is easily verified with simple algebra. In Figure 13.2c, let x, y, z, and w denote the areas of the differently shaded regions in the room. The area of the room is $x + y + z + w$, the combined area of the two carpets is $x + 2y + z$, and $x + y + z + w = x + 2y + z$ if and only if $y = w$. Thus we have proved.

The carpets theorem. *Place two carpets in a room. The area of the overlap equals the area of the uncovered area if and only if the combined area of the carpets equals the area of the floor.*

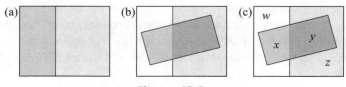

Figure 13.2.

The shapes of the room and the carpets are arbitrary, and the theorem holds for more than two carpets as long as at most two carpets overlap.

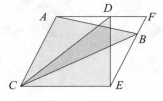

Figure 13.3.

As an example, see the parallelogram $ACEF$ in Figure 13.3. The area of triangles ABC and CDE are each one-half of the area of $ACEF$, so by the carpets theorem the area of the dark gray overlap quadrilateral equals the sum of the areas of the three white regions [Andreescu and Enescu, 2004].

For more applications of the carpets theorem, see Challenges 13.1 and 13.2.

13.2   The irrationality of $\sqrt{2}$ and $\sqrt{3}$

There are many proofs that $\sqrt{2}$ is irrational. Alexander Bogomolny's website *Mathematics Miscellany and Puzzles*, www.cut-the-knot.org, has more than 20. Most of them begin by assuming that $\sqrt{2}$ is rational, and then reach a contradiction, such as in one created by Stanley Tennenbaum [Conway, 2005] which we now present.

Assume $\sqrt{2}$ is rational and write $\sqrt{2} = m/n$, where m and n are positive integers and the fraction is in lowest terms. Then $m^2 = 2n^2$, so there exist two squares with integer sides m and n such that one has exactly twice the area of the other, and m and n are the smallest positive integers with this property. See Figure 13.4a.

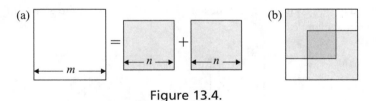

Figure 13.4.

Place the two small squares on the large one as shown in Figure 13.4b. By the carpets theorem, the area of the dark gray square equals the sum of the areas of the two white squares. But these squares also have integer sides $2n - m$ and $m - n$ smaller than m and n, respectively, a contradiction. Hence $\sqrt{2}$ is irrational.

Similarly, we can show that $\sqrt{3}$ is irrational using overlapping equilateral triangles. Let $T_s = s^2\sqrt{3}/4$ denote the area of an equilateral triangle with side s, and assume that $\sqrt{3}$ is rational with $\sqrt{3} = m/n$ in lowest terms, so that $m^2 = 3n^2$, or equivalently, $T_m = 3T_n$.

The side lengths of the dark gray and white triangles in Figure 13.5 are $2n - m$ and $2m - 3n$, respectively, so by the carpets theorem we have $T_{2m-3n} = 3T_{2n-m}$, or equivalently $(2m - 3n)^2 = 3(2n - m)^2$. Hence

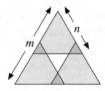

Figure 13.5.

$\sqrt{3} = (2m - 3n)/(2n - m)$, a contradiction since $0 < 2m - 3n < m$ and $0 < 2n - m < n$.

13.3 Another characterization of Pythagorean triples

In Section 7.4 and Challenge 7.4 we encountered characterizations of Pythagorean triples, triples (a, b, c) of integers such that $a^2 + b^2 = c^2$. The Pythagorean relation $a^2 + b^2 = c^2$ for a right triangle with sides a, b, and c suggests considering two square carpets with areas a^2 and b^2 in a square room with area c^2, as shown in Figure 13.6a [Teigen and Hadwin, 1971; Gomez, 2005]. By the carpets theorem, the area $(a + b - c)^2$ of the central dark gray square equals the sum $2(c - a)(c - b)$ of the areas of the two white rectangles.

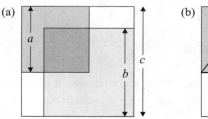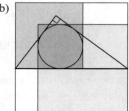

Figure 13.6.

Set $n = a + b - c$, $p = c - a$, and $q = c - b$. Since n, p, and q are integers and $a^2 + b^2 = c^2$ if and only if $n^2 = 2pq$, we have the following characterization: *there is a one-to-one correspondence between Pythagorean triples (a, b, c) and factorizations of even squares of the form $n^2 = 2pq$.* Furthermore, $a = n + q$, $b = n + p$, $c = n + p + q$, and (a, b, c) is primitive if and only if p and q are relatively prime. For example, $6^2 = 2 \cdot 1 \cdot 18$

corresponds to the triple (7, 24, 25), $6^2 = 2 \cdot 2 \cdot 9$ corresponds to (8, 15, 17), and $6^2 = 2 \cdot 3 \cdot 6$ corresponds to (9, 12, 15).

Finally, the side $a + b - c$ of the central square is the same as the diameter of the incircle (see Section 7.2) as illustrated in Figure 13.6b, and is less than the altitude to the hypotenuse.

Pythagorean oddities

In the bride's chair on the Greek stamp (Figure 1.1) and in the *Zhou bi suan jing* (Figure 2.1) we see the Pythagorean triple (3, 4, 5), the only one in which a, b, c are consecutive integers, and the prototype for triples in arithmetic progression. However, there are many triples in which one leg and the hypotenuse are consecutive, e.g., (5, 12, 13) and (7, 24, 25); and many for which the two legs are consecutive, e.g., (20, 21, 29) and (119, 120, 169). In 1643 Pierre de Fermat wrote a letter to Marin Mersenne asking for Pythagorean triples such that the both the hypotenuse and the sum of the legs are squares. There are infinitely many such triples, the smallest of which is (4565486027761, 1061652293520, 4687298610289). For details, see [Sierpiński, 1962].

13.4 Inequalities between means

Several nice inequalities for means can be derived from the overlapping squares icon illustrated at the beginning of this chapter. To do so, we relax the restriction that the sum of the areas of the overlapping squares equals the area of the enclosing square, as in the carpets theorem. See Figure 13.7.

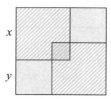

Figure 13.7.

When $x, y > 0$ two of the squares overlap unless $x = y$, and hence we have the inequality

$$2(x^2 + y^2) \geq (x + y)^2. \tag{13.1}$$

If we let $x = \sqrt{a}$ and $y = \sqrt{b}$ we have $2(a + b) \geq (\sqrt{a} + \sqrt{b})^2 = a + 2\sqrt{ab} + b$ which simplifies to the AM-GM inequality: for positive a and b,

$$\frac{a + b}{2} \geq \sqrt{ab}.$$

If we let $x = a/2$ and $y = b/2$ in (13.1) we have $(a^2 + b^2)/2 \geq ((a + b)/2)^2$, which on taking square roots yields the arithmetic mean-root mean square inequality: for positive a and b,

$$\sqrt{\frac{a^2 + b^2}{2}} \geq \frac{a + b}{2}.$$

If we let $x = 1/\sqrt{a}$ and $y = 1/\sqrt{b}$ in (13.1) we have $2\left(1/a + 1/b\right) \geq 1/a + 2/\sqrt{ab} + 1/b$, or equivalently, $(a + b)/ab \geq 2/\sqrt{ab}$. Taking reciprocals and multiplying by 2 yields the harmonic mean-geometric mean inequality: for positive a and b,

$$\sqrt{ab} \geq \frac{2ab}{a + b}.$$

We now generalize Figure 13.7 to overlapping rectangles within a rectangle.

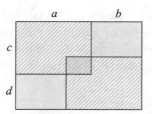

Figure 13.8.

When $a \geq b > 0$ and $c \geq d > 0$ (as shown in Figure 13.8), then $2(ac + bd) \geq (a + b)(c + d)$, or equivalently

$$ac + bd \geq ad + bc. \tag{13.2}$$

The inequality also holds for $b \geq a > 0$ and $d \geq c > 0$; it is reversed for $a \geq b > 0$ and $d \geq c > 0$, and for $b \geq a > 0$ and $c \geq d > 0$ and we have equality if and only if $a = b$ or $c = d$.

13.5 Chebyshev's inequality

As an application of (13.2) we can now prove *Chebyshev's inequality* (Pafnuty Lvovich Chebyshev, 1821–1894): *For any $n \geq 2$, let $0 < x_1 \leq x_2 \leq \cdots \leq x_n$. Then*

(i) *if $0 < y_1 \leq y_2 \leq \cdots \leq y_n$, then* $\sum_{i=1}^{n} x_i \sum_{j=1}^{n} y_j \leq n \sum_{i=1}^{n} x_i y_i$,

$$(13.3a)$$

(ii) *if $y_1 \geq y_2 \geq \cdots \geq y_n > 0$, then* $\sum_{i=1}^{n} x_i \sum_{j=1}^{n} y_j \geq n \sum_{i=1}^{n} x_i y_i$,

$$(13.3b)$$

with equality in each if and only if all the x_is are equal or all the y_is are equal.

For (13.3a), we use the result in (13.2) with $i \leq j$, $a = x_i$, $b = x_j$, $c = y_i$, and $d = y_j$, so $x_i y_j + x_j y_i \leq x_i y_i + x_j y_j$. Applying the inequality to the expansion of $(x_1 + x_2 + \cdots + x_n)(y_1 + y_2 + \cdots + y_n)$ yields a sum of terms of the form $x_i y_i$, establishing the result. For (13.3b), we do the same, but in this case the inequality is reversed since $b \geq a$ and $c \geq d$.

See Figure 13.9 for an illustration of the case $n = 4$ of (13.3a) with overlapping rectangles.

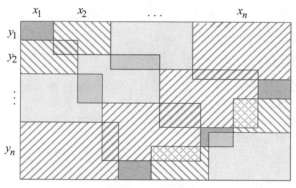

Figure 13.9.

13.6 Sums of cubes

An elegant proof [Golomb, 1965] of the formula for the sum of the cubes of the first n positive integers,

$$1^3 + 2^3 + 3^3 + \cdots + n^3 = (1 + 2 + 3 + \cdots + n)^2,$$

uses overlapping squares. We first express k^3 as k copies of k^2 for each k between 1 and n and then arrange the squares in a large square with sides $1 + 2 + \cdots + n$ as illustrated in Figure 13.10. When k is even two squares overlap, but the area of the overlap is the same as the area of a square (in white) not covered by shaded squares.

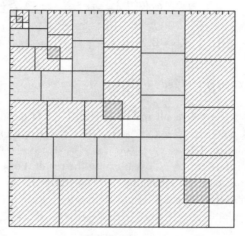

Figure 13.10.

13.7   Challenges

13.1. In the quadrilateral $ABCD$, let M, N, P, and Q be the midpoints of sides AB, BC, CD, and EF as shown in Figure 13.11. Show that the area of the dark gray quadrilateral equals the sum of the areas of the four white triangles.

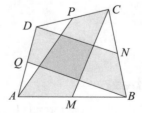

Figure 13.11.

13.2. In Figure 13.12 we see two rectangular carpets in a rectangular room. Prove that the area of the dark gray quadrilateral equals the sum of the areas of the six white triangles.

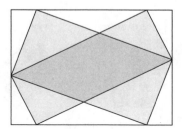

Figure 13.12.

13.3. *Inequalities by paper folding.* In Figures 13.13ab, we see paper cutouts consisting of shaded isosceles right triangles and a white rectangle and square. Folding the triangles on the dashed lines as indicated shows that the shaded area exceeds the white area in each case, establishing inequalities. What inequalities?

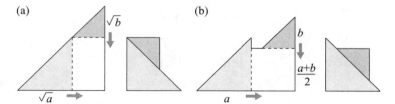

Figure 13.13.

13.4. *Cocircular points* [Gardner, 1975]. Five paper rectangles (one with a corner torn off) and six paper disks have been tossed on a table, as shown in Figure 13.14. Each corner of a rectangle and each place where edges are seen to meet marks a point. The problem is to find four sets of four cocircular points, that is, four points that lie on a circle. For example, the corners of the isolated rectangle in the bottom right of Figure 13.14 constitute such a set, since every rectangle possesses a circumcircle.

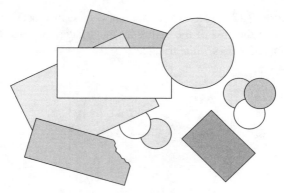

Figure 13.14.

Martin Gardner attributes this problem to Stephen Barr.

13.5. One magazine A lies on top of another one B, as illustrated in Figure 13.15. Does A cover more or less than half the area of B?

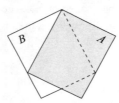

Figure 13.15.

CHAPTER 14

Yin and Yang

Its two elements unite and give rise to the concrete. Thus the multiplicity of things and human beings is produced. In their ceaseless successions the two elements of yin and yang constitute the great principles of the universe.

Zhang Zai (1020–1077)

In Chinese thought yin and yang represent the two great opposite but complementary creative forces at work in the universe. The yin and yang diagram, our icon, is known in Taoism as *taijitu* (diagram of the supreme ultimate), and enables us to visualize the philosophical idea of complementary opposites within a greater whole (e.g., day and night, feminine and masculine, good and evil, positive and negative, odd and even, etc.). The *taijitu*, a circle partitioned into two differently colored congruent parts by a curve consisting of two semicircles, appeared on the flag of the Kingdom of Korea in 1893, and is now on the flag of the Republic of Korea (South Korea). It has also been used on corporate logos, and the shape is popular with designers of jewelry, furniture, basins, and bowls.

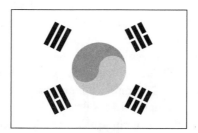

Flag of the Republic of Korea

Two copies of a two- or three-dimensional geometrical object may be combined to form a new figure whose area or volume is doubled (as with

173

yin and yang combining to form a circular disk). This simple idea has many nice applications in mathematics, which we explore after examining several properties of the icon.

14.1   The great monad

In 1917 Henry Ernest Dudeney (1857–1930), seen below in Figure 14.1, published his classic mathematical puzzles book, *Amusements in Mathematics*. Problem 158 concerns the yin and yang, which Dudeney calls *the great monad* (a *monad* is defined as a metaphysical entity, spiritual in nature, reflecting within itself the whole universe). In his problem Dudeney asks us to do two things:

1. divide the yin and the yang (see Figure 14.1a) into four pieces of the same size and shape by one cut and

2. divide the yin and the yang into four pieces of the same size by one straight cut.

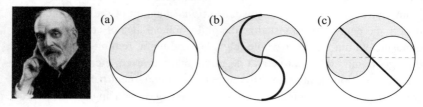

Figure 14.1.

While Dudeney did not make it explicit in his problem, it is clear that he intends us to solve the problem with the classical Euclidean tools, a compass and straightedge (which are all we will use for the constructions in this chapter).

To solve the first problem, we cut along the path indicated by the dark curve in Figure 14.1b, the boundary between yin and yang rotated 90°. The dark solid line that makes a 45° angle with the horizontal diameter (the gray dashed line), as shown in Figure 14.1c, solves the second problem. It bisects yin since the area of the gray semicircle below the dashed line is 1/8 the area of the entire circle, and the area of the gray circular sector above the dashed line is also 1/8 the area of the entire circle. Hence the semicircle and the sector together have 1/4 the area of the entire circle.

Other ways to bisect yin and yang exist as in Challenge 14.1.

Piecewise circular curves

The shared boundary of yin and yang is an example of a *piecewise circular curve*, a finite sequence of circular arcs with the endpoint of one arc coinciding with the beginning point of next [Banchoff and Giblin, 1994]. Piecewise circular curves are circular analogs of polygons. Other examples include the arbelos and salinon from Section 4.4, the lune from Challenge 4.7, the vesica piscis in Chapter 11, the Reuleaux triangle from Section 12.3, and the cardioid of Boscovich in Challenge 4.9.

In Figure 14.2 we see an asymmetrical yin and yang, generated by partitioning the horizontal diameter into two segments of lengths a and b. What is the ratio of the areas of the gray and white regions?

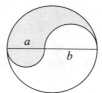

Figure 14.2.

The area of the gray region is

$$\frac{\pi}{8}a^2 + \frac{\pi}{8}(a+b)^2 - \frac{\pi}{8}b^2 = \frac{\pi}{4}a(a+b),$$

and similarly the area of the white region is $\pi b(a+b)/4$. Thus the ratio of the areas equals the ratio a/b of the two segments of the diameter.

If $b = 6a$, then the area of the gray region is $1/7$ the area of the circle. This raises the question: can we partition a circle into seven regions, each equal to $1/7$ the area of the circle? It is impossible with sectors of equal area, since that is equivalent to inscribing a regular heptagon in a circle but we can accomplish it with a yin and yang-type dissection.

Partition the diameter of a semicircle into seven intervals of equal width and construct six nested semicircles in the interior, as shown in Figure 14.3a. It is easy to show that the areas of the regions between the semicircles have

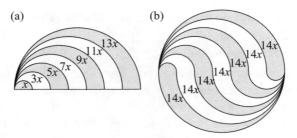

Figure 14.3.

the values indicated (the value of x is immaterial). Completing the circular disk as shown in Figure 14.3b shows that each of the regions has the same area, that is, $1/7$ the area of the circle.

14.2   Combinatorial yin and yang

The symmetry present in the yin and yang icon can be exploited to solve simple combinatorial problems. Perhaps the simplest (and best known) is the evaluation of $T_n = 1 + 2 + 3 + \cdots + n$. We represent it as a triangular array of balls as shown in Figure 14.4a, and call T_n the nth *triangular number*. Doubling T_n forms the rectangular array in Figure 14.4 that contains $n \times (n + 1)$ balls, thus

$$T_n = 1 + 2 + 3 + \cdots + n = \frac{n(n + 1)}{2}.$$

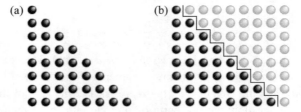

Figure 14.4.

The sum is an example of an *arithmetic progression*, a finite sequence in which the difference between two successive terms is a constant. If the first term in the sequence is a and the common difference is d, then the sum of the first n terms of the progression is

$$a + (a + d) + (a + 2d) + \cdots + [a + (n - 1)d] = \frac{n}{2}[2a + (n - 1)d].$$

That is, the sum equals one-half the number of terms times the sum of the first and last term in the progression. See Figure 14.5 [Conway and Guy, 1996], where we represent the terms in the progression by areas of rectangles.

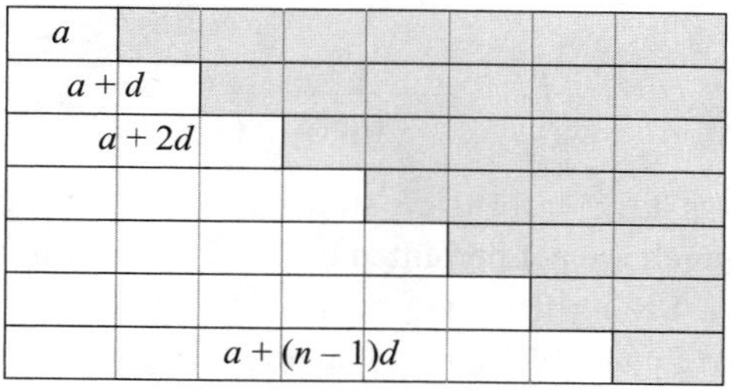

a							
	$a + d$						
		$a + 2d$					
		$a + (n - 1)d$					

Figure 14.5.

The yin and yang symmetry we have used in the plane can be applied in three dimensions where we duplicate an object to form a solid whose volume is easily calculated. For example, for $n \geq 1$ consider

$$S_n = \sum_{i=1}^{n} \sum_{j=1}^{n} (i + j - 1).$$

If we represent S_n by the collection of unit cubes in Figure 14.6a, then two copies of S_n fit into a rectangular box with dimensions $n \times n \times 2n$, as illustrated in Figure 14.6b. Hence $2S_n = 2n^3$, so that $S_n = n^3$.

(a)   (b)

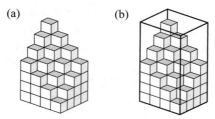

Figure 14.6.

A similar figure yields the generalized result

$$\sum_{i=1}^{m} \sum_{j=1}^{n} [a + (i - 1)b + (j - 1)c] = \frac{mn}{2}[2a + (m-1)b + (n-1)c].$$

As with one-dimensional arithmetic progressions, the sum is one-half the number of terms times the sum of the first $[(i, j) = (1, 1)]$ and last $[(i, j) = (m, n)]$ terms.

14.3 Integration via the symmetry of yin and yang

Problem A3 on the 1980 William Lowell Putnam Mathematical Competition asked competitors to evaluate

$$\int_0^{\pi/2} \frac{dx}{1 + (\tan x)^{\sqrt{2}}}.$$

The problem was difficult for many students. If they had access to graphing calculators (which are not permitted in the Competition) they may have had more success. Seeing that the graph of the integrand on the interval $[0, \pi/2]$ looks like Figure 14.7a, many students may have been able to exploit its symmetry to evaluate the integral.

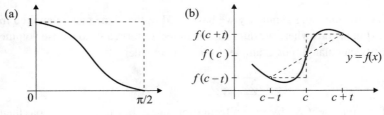

Figure 14.7.

The symmetry in the graph of the integrand is *symmetry with respect to a point*, the same symmetry exhibited by the boundary curve (the two semicircles) between yin and yang in Figure 14.1a. To be precise, the graph of $y = f(x)$ is symmetric with respect to the point $(c, f(c))$ if $f(c - t) + f(c + t) = 2f(c)$ when $c, c - t$, and $c + t$ are in the domain of the function. See Figure 14.7b.

Suppose that f is continuous on the interval $[a, b]$ and that the graph of f is symmetric with respect to the point whose x-coordinate is the midpoint $(a + b)/2$ of $[a, b]$, so f satisfies $f(x) + f(a + b - x) = 2f((a + b)/2)$ for all x in $[a, b]$. Such functions are easy to integrate:

$$\int_a^b f(x)\, dx = (b - a)f\left(\frac{a + b}{2}\right) = \frac{1}{2}(b - a)[f(a) + f(b)].$$

While an analytic proof is straightforward, perhaps Figure 14.8 is more instructive, where we see a rectangular version of yin and yang.

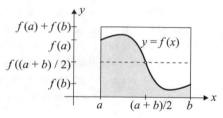

Figure 14.8.

Since the integrand in the Putnam problem satisfies $f(x) + f(\pi/2-x) = 2f(\pi/4) = 1$ on $[0, \pi/2]$, the solution is $(1/2)(\pi/2)[1 + 0] = \pi/4$. A similar problem appeared as Problem B1 on the 1987 Putnam Competition: Evaluate

$$\int_2^4 \frac{\sqrt{\ln(9 - x)} \, dx}{\sqrt{\ln(9 - x)} + \sqrt{\ln(x + 3)}}.$$

The same procedure can be used to show that the answer is 1. The symbolic algebra program *Mathematica* (v. 7.01) is unable to evaluate the two Putnam integrals. See Challenge 14.6 for other integrals that can be evaluated by exploiting symmetry.

14.4 Recreational yin and yang

In Section 14.1 we saw Dudeney's great monad puzzle (also see Challenge 14.1). The yin and yang in Figure 14.1a not only partitions the circular disk into two congruent regions, but exhibits interesting rotational symmetry. Applying the symmetry found in yin and yang is a common procedure in the world of recreational mathematics.

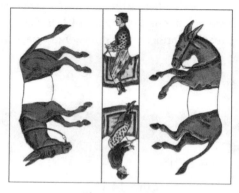

Figure 14.9.

In 1871 Sam Loyd created a puzzle he called the "trick donkeys." The following year it was marketed by P. T. Barnum under the name "P. T. Barnum's Trick Mules." Millions of copies were printed on cards and sold, which made Sam Loyd a rich man within a year. The puzzle is shown in Figure 14.9. Its object is to cut the card into three (and only three) pieces along the two vertical lines and then rearrange them so that each rider appears to be riding a mule.

The solution is to arrange the two pieces with the mules back to back as shown in Figure 14.10a (exhibiting a form of yin and yang symmetry), and then place the piece with the riders on top as shown in Figure 14.10b.

(a)   (b)

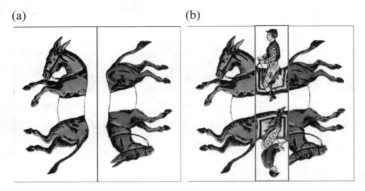

Figure 14.10.

The game of Daisy is a children's game for two played with n counters arranged on a circular board, as shown in Figure 14.11a for $n = 9$. Alternating turns, players remove either one or two counters but for two they must be adjacent. The player removing the last marker wins. After each player has had one turn, the board might look like Figure 14.11b. You may wish to play a few games before reading further to see if one of the players has a winning strategy.

(a)   (b)

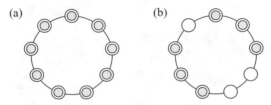

Figure 14.11.

Did you discover that the second player can always win? When n is even, the second player mimics the first player's move, removing one or two counters diametrically opposite the counters removed by the first player. When n is odd, if the first player removes one counter on the first turn, the second player removes the two diametrically opposite. If the first player removes two counters on the first turn, the second player removes one diametrically opposite. The board will then resemble Figure 14.11b and the second player plays the strategy for n even.

14.5 Challenges

14.1. Show that the cuts illustrated in Figure 14.12 also bisect yin and yang from Figure 14.1a. [Hints: If the radius of the monad is 1, then the radius of the circular cut in (b) is $\sqrt{2}/2$, and in (c) the radii of the semicircles that make up the cut are $\phi/2$ and $1/2\phi$ where ϕ is the golden ratio, $\phi = (1 + \sqrt{5})/2$.]

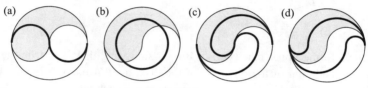

Figure 14.12.

14.2. Use two copies of the array of balls in Figure 14.13 to show that

$$1 + 3 + 5 + \cdots + (2n - 1) = n^2.$$

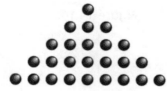

Figure 14.13.

14.3. In Section 4.2 we showed that $T_n = n(n + 1)/2$, that is, the nth triangular number is the binomial coefficient $\binom{n+1}{2}$. Since $\binom{k}{2}$ is the number of ways to chose two elements from a set of k elements, there should exist a one-to-one correspondence between a set of T_n elements and the set of two-element subsets of another set with $n + 1$ elements. Find one.

14.4. Prove that every integer $N > 1$ that is not a power of two can be expressed as a sum of consecutive integers.

14.5. Use an array of cubes (similar to the array in Figure 14.6) to illustrate the formula for the sum of squares of the first n positive integers:

$$1^2 + 2^2 + 3^2 + \cdots + n^2 = \frac{n(n + 1)(2n + 1)}{6}.$$

[You may need to use six copies!]

14.6. Evaluate the integrals:

(a) $\int_{-1}^{1} \arctan(e^x)\, dx$ (b) $\int_{0}^{\pi/4} \ln(1 + \tan x)\, dx$

(c) $\int_{0}^{2} \frac{dx}{x + \sqrt{x^2 - 2x + 2}}$ (d) $\int_{0}^{2} \sqrt{x^2 - x + 1} - \sqrt{x^2 - 3x + 3}\, dx$

(e) $\int_{0}^{4} \frac{dx}{4 + 2^x}$ (f) $\int_{0}^{2\pi} \frac{dx}{1 + e^{\sin x}}$

14.7. In the upper half (yin) of the yin and yang disk, let A be the right endpoint of the diameter of the disk, as illustrated in Figure 14.14. Find all points B in the boundary of yin such that there exists a point C in the boundary of yin for which ABC is a right triangle with the right angle at B.

Figure 14.14.

CHAPTER **15**

Polygonal Lines

> It is known that no Circle is really a Circle, but only a Polygon with a very large number of very small sides. As the number of the sides increases, a Polygon approximates to a Circle; and, when the number is very great indeed, say for example three or four hundred, it is extremely difficult for the most delicate touch to feel any polygonal angles.
>
> Edwin Abbott Abbott
> *Flatland* (1884)

Informally, a polygonal line is a collection of line segments where adjacent segments share endpoints. More formally we have the following definition. Given a finite sequence $\{P_0, P_1, P_2, \ldots, P_n\}$ of $n + 1$ distinct points in the plane, called *vertices*, a *polygonal line* consists of the vertices and the associated *edges*, the line segments $P_0 P_1, P_1 P_2, \ldots, P_{n-1} P_n$.

Examples of polygonal lines in man-made objects include the folding wooden ruler, the data display, and the articulated ladder in Figure 15.1.

(a) (b) (c)

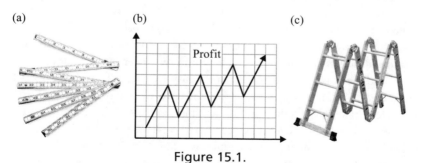

Figure 15.1.

When $P_n = P_0$ we have a closed figure, called a *polygon* (or an *n-gon* when we wish to indicate number of edges). We have encountered

183

polygons in many of the preceding chapters—triangles and squares in Chapter 1, trapezoids in Chapter 2, and so on.

We begin with a simple question: how do we draw the simplest polygonal line, a straight line segment? Next we derive formulas for polygonal numbers by considering the vertices of a polygonal line. Then we note the use of polygonal lines in calculus. After deriving some general results about convex polygons, we conclude by using regular polygons to obtain some results about cycloids and cardioids.

15.1   Lines and line segments

The simplest polygonal lines are segments of a straight line. But what is meant by "straight line?" In his dialog *Parmenides*, Plato writes "straight is that of which the middle is in front of both extremities." In the *Elements* of Euclid, Definition I.4 states "A straight line lies equally with respect to the points on itself." Not surprisingly, Euclid never makes use of this definition.

Setting aside the problem of defining a straight line or line segment, how do we *draw* one? That seems easy: use a straightedge or ruler. But how do we know that the straightedge is straight?

When we draw a circle, we usually do not trace a circular disk with a pencil or pen, we use (or used to use, before computer software existed) a compass. The compass is a mechanical device that enables us to implement the definition of the circle as the locus of points equidistant from a fixed point. Euclid's definition of a straight line does not help much with drawing a line.

Is it possible to create a mechanical device for drawing a straight line, analogous to the compass for drawing circles? This was an important question in the 19th century, when many mechanical devices were invented, motivated by the industrial revolution. Linkages (rigid bars made of metal or wood joined by rivets at their ends) were created to turn circular motion into linear motion and vice-versa. The Scottish engineer James Watt (1736–1819) and the Russian mathematician Pafnuty Lvovich Chebyshev (1821–1894) created linkages to produce approximately linear motion, but the first to create a linkage to produce truly linear motion from circular motion was one designed by the French engineer Charles-Nicolas Peaucellier (1832–1913) in 1864, and independently rediscovered by the Russian mathematician Lippman Lipkin (1851–1875) in 1871. This linkage is variously known as the *Peaucellier-Lipkin linkage*, the *Peaucellier cell*, or the *Peaucellier inversor*.

In 1876 Alfred Bray Kempe gave a lecture at London's South Kensington Museum with the title *How to Draw a Straight Line*, which was published as a small book the following year [Kempe, 1877]. In Figure 15.2 we see Kempe's illustration of the Peaucellier-Lipkin linkage.

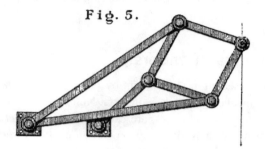

Fig. 5.

Figure 15.2.

In his lecture and book, Kempe presented a sophisticated argument showing that the linkage describes a straight line. Our argument is simpler, based on the law of cosines. In Figure 15.3, the lengths of the parts of the linkage are $|BC| = |BD| = a$, $|AC| = |CP| = |AD| = |DP| = b$ with $a > b > 0$, and $|AE| = |BE| = r$. Since B and E are fixed points, A will be on a circle of radius r centered at E. Let $\angle ABE = \alpha$, $\angle CAP = \beta$, and let Q be the foot of the perpendicular from P to the x-axis.

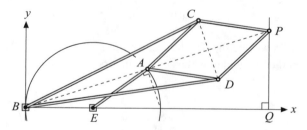

Figure 15.3.

To show that P describes a straight line, we show that its x-coordinate depends only on a, b, and r, and not on α or β. Since $|AB| = 2r \cos \alpha$ and $|AP| = 2b \cos \beta$, the x-coordinate of P is $|BQ| = |BP| \cos \alpha = (2r \cos \alpha + 2b \cos \beta) \cos \alpha$. Applying the law of cosines to triangle ABC yields

$$a^2 = b^2 + (2r \cos \alpha)^2 - 2b(2r \cos \alpha) \cos(\pi - \beta),$$

whence

$$\frac{a^2 - b^2}{2r} = (2r \cos \alpha + 2b \cos \beta) \cos \alpha = |BQ|.$$

Thus the x-coordinate of P depends only on a, b, and r, so that P describes a straight line as claimed.

15.2   Polygonal numbers

The physical representation of numbers by objects such as pebbles dates back at least to the ancient Greek geometers. Numbers with such representations, such as squares and cubes, are called *figurate numbers*. When the representation is in the shape of a polygon we call it a *polygonal number*.

The simplest of the polygonal numbers are the triangular and square numbers. The first five triangular numbers—1, 3, 6, 10, and 15—are illustrated in Figure 15.4. In Section 14.2 we saw that the nth triangular number $T_n = n(n+1)/2$.

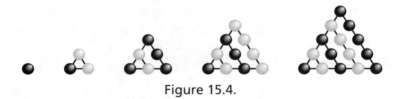

Figure 15.4.

In Figure 15.5 we see the first five square numbers $S_n = n^2$. The figure illustrates that the nth square is the sum of the first n odd numbers.

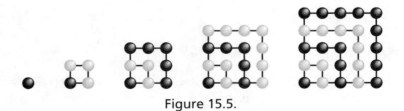

Figure 15.5.

The first four *pentagonal numbers*—1, 5, 12, and 22—are illustrated in Figure 15.6.

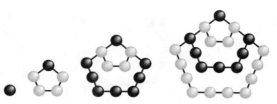

Figure 15.6.

To find a formula for P_n, the nth pentagonal number, we exploit relationships between pentagonal and triangular numbers. In Figure 15.7a we distort the pentagon to a trapezoid showing that $P_n = T_{2n-1} - T_{n-1}$, while in Figure 15.7b, we have $P_n = (1/3)T_{3n-1}$, both of which yield $P_n = n(3n-1)/2$.

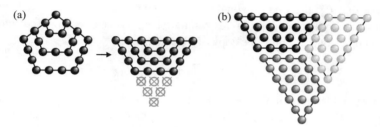

Figure 15.7.

The nth *hexagonal number* H_n is defined similarly. In Figure 15.8a we see H_n for $n = 4$ (i.e., $H_4 = 28$) and in Figures 15.8bcd we see that every hexagonal number is a triangular number and that their common value is the product of the subscripts (here $H_4 = T_7 = 4 \cdot 7$). In general we have $H_n = T_{2n-1} = n(2n-1)$.

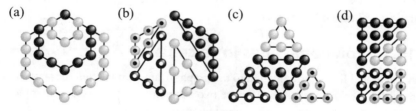

Figure 15.8.

Just as every pentagonal number is the difference of two triangular numbers, every *octagonal number* O_n is the difference of two squares. In Figure 15.9 we see $O_n = (2n-1)^2 - (n-1)^2$ illustrated for $n = 4$. Hence $O_n = n(3n-2)$.

Figure 15.9.

The relationship between a polygonal number and the triangular numbers can be exploited to find the formula for P_n^k, the nth k-gonal number for any integers $n \geq 1$ and $k \geq 3$ (e.g., $P_n^3 = T_n$, $P_n^4 = S_n$, etc.). In Figure 15.8b we saw that P_4^6, the 4th hexagonal number, could be decomposed into three copies of T_3 plus one copy of T_4. Similarly, P_n^k can be decomposed into $k-3$ copies of T_{n-1} plus one copy of T_n, and hence

$$P_n^k = T_n + (k-3)T_{n-1} = \frac{n}{2}[(k-2)n - (k-4)]. \qquad (15.1)$$

Polygonal numbers have long played a role in the theory of numbers. Their properties were investigated by Nicomachus of Gerasa (circa 100) and Diophantus of Alexandria (circa 250). In 1638 Pierre de Fermat (1601–1665) wrote that every positive integer is the sum of at most three triangular numbers, at most four squares, and in general, at most n n-gonal numbers. His proof, if it existed, has never been found. Carl Friedrich Gauss (1777–1855) proved the triangular case, and on July 10, 1796 he wrote in his diary "EYPHKA! Num $= \Delta + \Delta + \Delta$." The case for square numbers was proven in 1770 by Joseph Louis Lagrange (1736–1813) and is known as Lagrange's four-squares theorem. In 1813 Augustin-Louis Cauchy (1789–1857) proved the general case of Fermat's claim.

15.3   Polygonal lines in calculus

A polygonal line approximation to the graph of function is often the starting point for results in calculus. Here are two examples. In each we consider the function $y = f(x)$ on some specified interval.

1. *The trapezoidal rule.* To approximate the value of $\int_a^b f(x)dx$ (with $a < b$ and $f(x) \geq 0$), we appeal to the area interpretation of the integral and replace the graph of $y = f(x)$ on the interval $[a, b]$ with a polygonal line with vertices $\{P_0, P_1, P_2, \ldots, P_n\}$ where $P_i = (x_i, y_i)$ with $\Delta x = (b-a)/n$, $x_i = a + i\Delta x$, and $y_i = f(x_i)$. The area under

the graph of the polygonal line is the sum of the areas of n trapezoids, which yields the approximation

$$\int_a^b f(x)dx \approx \frac{\Delta x}{2}(y_0 + 2y_1 + 2y_2 + \cdots + 2y_{n-1} + y_n).$$

2. *Arc length.* The development of the integral formula for the length of the graph of $y = f(x)$ on the interval $[a, b]$ begins with $\sum_{i=1}^n |P_{i-1} P_i|$, the total length of the same polygonal line approximation to the graph of $y = f(x)$ used in the derivation of the trapezoidal rule.

15.4 Convex polygons

A polygon is *convex* if every line segment joining two points of the polygon lies within the polygon. Consequently its interior angles measure less than $180°$, and its diagonals (the line segments joining non-adjacent vertices) lie within it. Properties of the angles and diagonals of a convex n-gon include the following:

(i) *the sum of the angle measures is* $(n-2)180°$

(ii) *there are* $n(n-3)/2$ *diagonals*

(iii) *the diagonals intersect in at most* $\binom{n}{4}$ *interior points.*

Since the n-gon is convex, we can select one vertex and draw $n-3$ diagonals connecting it to all the other vertices except the two neighboring vertices. The $n-3$ diagonals divide the n-gon into $n-2$ triangles, so the sum of the interior angles is $(n-2)180°$. For (ii), $n-3$ diagonals terminate at each vertex, so the total number of endpoints of diagonals is $n(n-3)$. Since each diagonal has two endpoints, there are $n(n-3)/2$ diagonals. For (iii), each interior point of intersection of two diagonals is also the point of intersection of the diagonals of at least one quadrilateral whose four vertices are vertices of the n-gon (see Figure 15.10), and four vertices of the n-gon can be chosen in $\binom{n}{4}$ ways.

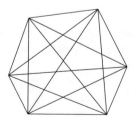

Figure 15.10.

Consider an n-gon and all of its diagonals. How many triangles can be found? We answer the question when no three diagonals meet at an interior point, as in Figure 15.10. Let P be a convex n-gon with the property that no three diagonals meet at an interior point. Then the number of triangles in P whose vertices are either interior points or vertices of P is

$$\binom{n}{3} + 4\binom{n}{4} + 5\binom{n}{5} + \binom{n}{6}.$$

Following [Conway and Guy, 1996], we count the triangles according to the number of vertices they share with P. There are $\binom{n}{3}$ triangles with all three vertices in common with P (see Figure 15.11a), $4\binom{n}{4}$ triangles that have exactly two vertices in common with P (see Figure 15.11b), $5\binom{n}{5}$ triangles that have exactly one vertex in common with P (see Figure 15.11c), and $\binom{n}{6}$ triangles all of whose vertices are interior points of P.

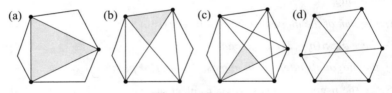

Figure 15.11.

A similar question is the following: into how many regions do the diagonals of a convex n-gon P partition the interior of the polygon? See [Honsberger, 1973; Freeman, 1976; Alsina and Nelsen, 2010] for proofs that if P has the property that no three diagonals meet at an interior point, then the number of regions in the resulting partition of P is

$$\binom{n}{4} + \binom{n-1}{2}.$$

Triangulating a polygon means partitioning it into non-overlapping triangles with nonintersecting diagonals that join pairs of vertices. For example, the number of triangulations of a square, a pentagon, and a hexagon are 2, 5, and 14, respectively. See Figure 15.12.

The problem of counting the number of triangulations of a convex n-gon has a distinguished history. The problem (and tentative answers for some n)

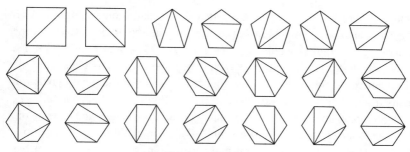

Figure 15.12.

was stated by Leonhard Euler (1707–1783) in a letter to Christian Goldbach (1690–1764) [Euler, 1965]. Euler also communicated the problem to Jan Andrej Segner (1704–1777), who solved it. Later it was presented as an open challenge by Joseph Liouville (1809–1894) and solved by several mathematicians, including Gabriel Lamé (1795–1870). In this section we present Lamé's combinatorial arguments [Lamé, 1838].

Let $T_n, n \geq 3$, denote the number of triangulations of a convex n-gon. Then $T_3 = 1$ *and*

(i) $T_{n+1} = T_n + T_3 T_{n-1} + T_4 T_{n-2} + \cdots + T_{n-2} T_4 + T_{n-1} T_3 + T_n$ *for* $n \geq 3$,

(ii) $T_n = n (T_3 T_{n-1} + T_4 T_{n-2} + \cdots + T_{n-2} T_4 + T_{n-1} T_3)/(2n - 6)$ *for* $n \geq 4$, *and*

(iii) $T_{n+1} = (4n - 6)T_n/n$ *for* $n \geq 3$.

We illustrate the proof of (i) with an octagon, but it is clear how the proof proceeds for an n-gon. See Figure 15.13.

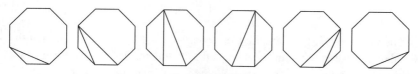

Figure 15.13.

Consider the bottom side of the octagon. After triangulation, it is the base of exactly six triangles, as illustrated in the figure. In each case there is a polygon (possibly just a line segment or a "2-gon") to the left and one to the right of the triangle that must also be triangulated.

In the first octagon in Figure 15.13, there is a line segment on the left and a 7-gon on the right, which can be triangulated in T_7 ways. In the second octagon there is a triangle on the left and a hexagon on the right, which together can be triangulated in $T_3 T_6$ ways. Continuing in this fashion for the remaining four figures yields $T_8 = T_7 + T_3 T_6 + T_4 T_5 + T_5 T_4 + T_6 T_3 + T_7$.

We illustrate the proof of (ii) with a heptagon, but it is clear how the proof proceeds for an n-gon. See Figure 15.14. Each diagonal appears in many triangulations. The leftmost diagonal in Figure 15.14 appears in $T_3 T_6$ triangulations, as it partitions the heptagon into a triangle and a hexagon. The diagonal to its right appears in $T_4 T_5$ triangulations, as it partitions the heptagon into a quadrilateral and a pentagon. Since the heptagon has seven vertices, the quantity $L_7 = 7 (T_3 T_6 + T_4 T_5 + T_5 T_4 + T_6 T_3)$ includes all possible triangulations counted by means of the diagonals, but many of them have been counted several times. In general, we have $L_n = n (T_3 T_{n-1} + T_4 T_{n-2} + \cdots + T_{n-1} T_3)$.

Figure 15.14.

Each triangulation of an n-gon has $n - 3$ different diagonals, and any set of $n - 3$ different diagonals will appear in the sum L_n exactly $2(n - 3)$ times, since each diagonal has two endpoints. Thus $T_n = L_n / (2n - 6)$, as claimed. Finally, (iii) is an immediate consequence of (i) and (ii), and (iii) provides the induction step for the proof than an explicit formula for T_n is

$$T_n = \frac{1}{n - 1} \binom{2n - 4}{n - 2}.$$

The numbers $\{T_n\}_{n=3}^{\infty}$ are the *Catalan numbers* $C_n = T_{n+2}$, i.e., the nth Catalan number is the number of ways to triangulate a $(n+2)$-gon for $n \geq 1$. They appear in the solutions to many other combinatorial problems, such as parentheses in products, diagonal-avoiding paths in a square lattice, and binary trees. They are named after the Belgian mathematician Eugène Charles Catalan (1814–1894).

15.5 Polygonal cycloids

It is well known that when a circle rolls along a line, the curve generated by a point on the circle is the *cycloid*. Two of the nice properties of a cycloid are:

(i) the area under a cycloid is three times the area of the generating circle; and (ii) the length of a cycloid is four times the diameter of the generating circle. Do similar results hold if the circle is replaced by a regular polygon?

The Helen of Geometers

The cycloid was first studied by Nicolas of Cusa (1401–1464) and Charles de Bouvelles (1471–1553), but the name "cycloid" is attributed to Galileo Galilei (1564–1642), who also advocated its use as the shape of an arch in architecture.

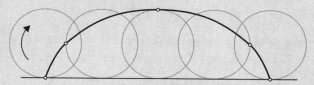

Figure 15.15.

Gilles Personne de Roberval (1602–1675) determined the area under one arch, and Christopher Wren (1632–1723) found its length. The cycloid became known as "the Helen of geometers" for the many quarrels over priority it provoked during the 17th century. Christiaan Huygens (1629–1695) showed that an inverted cycloid solves the tautochrone problem—to find the curve down which a ball placed anywhere will roll to the bottom in the same amount of time. The solution to the tautochrone problem makes an appearance in the following passage from Herman Melville's novel *Moby Dick* (1851):

"It was in the left hand try-pot of the *Pequod*, with the soapstone diligently circling round me, that I was first indirectly struck by the remarkable fact, that in geometry all bodies gliding along the cycloid, my soapstone for example, will descend from any point in precisely the same time."

If we replace the circle by a regular polygon, the curve generated by a vertex of the polygon consists of arcs of circles, sometimes called a *cyclogon* [Apostol and Mnatsakanian, 1999]. See Figure 15.16a for an octagon rolling along a line.

(a) 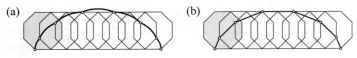 (b)

Figure 15.16.

If we replace the arcs by their chords, the resulting figure is called a *polygonal cycloid*, and is illustrated for the octagon in Figure 15.16b. We find the area under one arch of a polygonal cycloid and its length. First we prove a lemma of interest in its own right [Ouellette and Bennett, 1979]:

If $V_1, V_2, \ldots, V_n$ are the vertices of a regular n-gon with circumradius R, and if P is a point on the circumcircle of the n-gon, then

$$|PV_1|^2 + |PV_2|^2 + \cdots + |PV_n|^2 = 2nR^2.$$

Place the n-gon in the xy-plane so that the center of the circumcircle is at the origin and let $V_i = (a_i, b_i)$ and $P = (u, v)$. Then

$$|PV_1|^2 + |PV_2|^2 + \cdots + |PV_n|^2 = \sum_1^n (u - a_i)^2 + \sum_1^n (v - b_i)^2$$

$$= n(u^2 + v^2) - 2u \sum_1^n a_i - 2v \sum_1^n b_i + \sum_1^n (a_i^2 + b_i^2)$$

$$= 2nR^2 - 2u \sum_1^n a_i - 2v \sum_1^n b_i,$$

since $u^2 + v^2 = R^2$ and $a_i^2 + b_i^2 = R^2$. To complete the proof we need only show that $\sum_1^n a_i = \sum_1^n b_i = 0$. Place equal weights at the vertices of the n-gon. Since their center of gravity will be the center of the circumcircle, the x and y moments of the system are zero, hence $\sum_1^n a_i = \sum_1^n b_i = 0$. (A formal proof using complex numbers can be found in [Ouellette and Bennett, 1979].)

A special case of interest is when P is one of the vertices of the n-gon. Then the sum of the squared distances from one vertex of a regular n-gon with circumradius R to each of the other $n - 1$ vertices is $2nR^2$.

We now prove that the area property for the cycloid also holds for polygonal cycloids: *When a regular polygon is rolled along a line, the area of the polygonal cycloid generated by a vertex of the polygon is three times the area of the polygon.*

Let R denote the circumradius of the n-gon, and let $d_1, d_2, \ldots, d_{n-1}$ denote the distances from a vertex (say V_1) of the n-gon to the other $n - 1$ vertices, as illustrated in Figure 15.17a.

Let A be the numerical value of the area of the regular n-gon with circumradius 1. Figure 15.17b shows the polygonal cycloid generated by V_1 as the

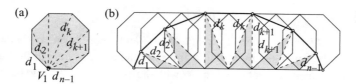

Figure 15.17.

polygon rolls along the line. The region under the polygonal cycloid can be partitioned into $n - 2$ shaded triangles and $n - 1$ white isosceles triangles, each with apex angle $2\pi/n$ (we measure angles in radians since we use trigonometric functions of angles in this section and the next). The equal-length sides in the isosceles triangles are successively $d_1, d_2, \ldots, d_{n-1}$. The $n - 2$ shaded triangles are congruent to the $n - 2$ triangles formed by the diagonals in the n-gon, and hence their areas sum to $R^2 A$. The sum of the areas of the $n - 1$ white isosceles triangles is

$$\frac{1}{n} A \big(d_1^2 + d_2^2 + \cdots + d_{n-1}^2 \big) = \frac{A}{n} \cdot 2nR^2 = 2R^2 A.$$

Hence the area under the polygonal cycloid is $3R^2 A$, or three times the area of the generating n-gon. Taking the limit as the number of sides of the n-gon goes to infinity shows that the area under one arch of the cycloid is three times the area of the generating circle.

We now turn our attention to the length of a polygonal cycloid, and prove *When a regular polygon is rolled along a line, the length of the polygonal cycloid generated by a vertex of the polygon is four times the sum of the inradius and the circumradius of the polygon.*

Let r and R denote, respectively, the inradius and circumradius of the n-gon, and let L_k denote the length of the segment of the polygonal cycloid in the isosceles triangle with equal sides d_k in Figure 15.17b. Since $d_k = 2R \sin(k\pi/n)$, we have

$$L_k = 4R \sin \frac{k\pi}{n} \sin \frac{\pi}{n} = 2R \left[\cos \frac{(k-1)\pi}{n} - \cos \frac{(k+1)\pi}{n} \right]$$

and hence the length of the polygonal cycloid is

$$\sum_{k=1}^{n-1} L_k = 2R \sum_{k=1}^{n-1} \left[\cos \frac{(k-1)\pi}{n} - \cos \frac{(k+1)\pi}{n} \right]$$
$$= 4R \left(1 + \cos \frac{\pi}{n} \right) = 4R + 4r,$$

where the final step follows from $r = R \cos(\pi/n)$. Consequently the length of one arch of a cycloid is four times the diameter of the generating circle.

15.6   Polygonal cardioids

The *cardioid* (from the Greek καρδια "heart" and ειδοs "form" or "appearance") can be generated by rolling a circle around a fixed circle of the same size, as illustrated in Figure 15.18a. As with the cycloid, the cardioid has two nice properties: (i) the area of a cardioid is six times the area of the generating circle; and (ii) the length of a cardioid is eight times the diameter of the generating circle. It is probably not surprising that similar results hold if the circles are replaced by regular polygons, as illustrated in Figure 15.18b in the case of the octagon.

(a)   (b)

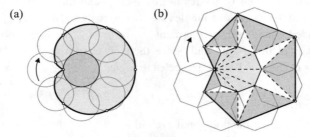

Figure 15.18.

Although cardioid means "heart-shaped," many observers think it looks more like the cross-section of a peach, a plum, or a tomato. Nonetheless, it is associated with the familiar heart icon that appears on Valentine's Day cards and gifts, playing cards, and Milton Glasser's "I♥NY" logo. The cardioid appears in a similar logo expressing affection for mathematics, as seen in Figure 15.19.

Figure 15.19.

Our work to find the area enclosed by the polygonal cardioid in Figure 15.18b parallels that used to find the area of a polygonal cycloid. Let A be the numerical value of the area of the regular n-gon with circumradius 1. The region enclosed by the polygonal cardioid can be partitioned into $n - 2$ white triangles and $n - 1$ dark gray isosceles triangles, each with apex angle $4\pi/n$. The equal-length sides in the isosceles triangles are $d_1, d_2, \ldots, d_{n-1}$. The $n - 2$ white triangles are congruent to the $n - 2$ triangles formed by the diagonals in the original n-gon, and hence their areas sum to $R^2 A$. The sum of the areas of the $n - 1$ dark gray isosceles triangles is

$$
\frac{1}{2}\left(d_1^2 + d_2^2 + \cdots + d_{n-1}^2\right) \sin \frac{4\pi}{n} = \left(d_1^2 + d_2^2 + \cdots + d_{n-1}^2\right) \sin \frac{2\pi}{n} \cos \frac{2\pi}{n}
$$

$$
= \frac{2A}{n}\left(d_1^2 + d_2^2 + \cdots + d_{n-1}^2\right) \cos \frac{2\pi}{n}
$$

$$
= \frac{2A}{n} \cdot 2nR^2 \cos \frac{2\pi}{n} = 4R^2 A \cos \frac{2\pi}{n}.
$$

Hence the area enclosed by the polygonal cardioid is $[2 + 4\cos(2\pi/n)]R^2 A$, or almost six times the area of the generating n-gon for large n. Taking the limit as n goes to infinity shows that the area enclosed by the cardioid is six times the area of the generating circle.

In Challenge 15.6 you will show that the length of the polygonal cardioid generated by an n-gon is $8R \cos^2(\pi/n) + 8r$, or almost eight times the sum of the inradius r and circumradius R of the n-gon. Consequently, the length of a cardioid is eight times the diameter of the generating circle.

There are many ways to define a cardioid. The point where the cardioid intersects the fixed circle (with radius a) is called the *cusp* of the cardioid, and if we introduce a polar coordinate system with the pole at the cusp and the polar axis as the extension of the diameter of the fixed circle through the cusp, then the center of the circle is $(a, 0)$ and the equation of the cardioid is $r = 2a(1 + \cos \theta)$. See [Pedoe, 1976] for details.

The Mandelbrot cardioid

The *Mandelbrot set* (the dark region in Figure 15.20) is the set of complex numbers c such that the limit of the sequence $\{c, c^2 + c, (c^2 + c)^2 + c, \ldots\}$ is *not* ∞. The boundary of the central bulb of the Mandelbrot set appears to be a cardioid. And so it is—for a proof, see [Branner, 1989].

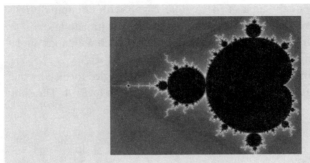

Figure 15.20.

15.7 Challenges

15.1. A number that can be written as the sum of two or more consecutive positive integers is called a *polite number*. For example, every pentagonal number is polite, as seen in Figure 15.7a. In Challenge 14.4 you proved that if n is not a power of 2, then n is polite. Now prove the converse: if n is polite, then n is not a power of 2. (The polite numbers that are not triangular are often called *trapezoidal numbers*).

15.2. If P_n^k represents the nth k-gonal number, show that $P_n^{k+1} = P_n^k + T_{n-1}$ and $P_{n+1}^k = P_n^k + (k-2)n + 1$ for $k \geq 3$ and $n \geq 2$. As a consequence it is easy to construct a table of polygonal numbers, beginning with $P_1^k = 1$:

$n \rightarrow$	1	2	3	4	5	6	7	8...
Triangular	1	3	6	10	15	21	28	36
Square	1	4	9	16	25	36	49	64
Pentagonal	1	5	12	22	35	51	70	92
Hexagonal	1	6	15	28	45	66	91	120
Heptagonal	1	7	18	34	55	81	112	148
Octagonal	1	8	21	40	65	96	133	176
etc.								

15.3. If T_k is the kth triangular number (with $T_0 = 0$), $0 \leq a, b, c \leq n$, and $2n \leq a + b + c$ show that

$$T_a + T_b + T_c - T_{a+b-n} - T_{b+c-n} - T_{c+a-n} + T_{a+b+c-2n} = T_n.$$

(While an algebraic proof is possible, consider a triangular array of T_n objects.) Some special cases are of interest:

(a) $(n; a, b, c) = (2k - j; k, k, k)$: $3(T_k - T_j) = T_{2k-j} - T_{2j-k}$;

(b) $(n; a, b, c) = (a + b + c; 2a, 2b, 2c)$: $T_{a+b+c} + T_{a+b-c} + T_{a-b+c} + T_{-a+b+c} = T_{2a} + T_{2b} + T_{2c}$;

(c) $(n; a, b, c) = (3k; 2k, 2k, 2k)$: $3(T_{2k} - T_k) = T_{3k}$.

15.4. Let P be a regular n-gon with circumradius R. Show that the sum of the squares of the lengths of all the sides and all the diagonals of P is $n^2 R^2$.

15.5. What are the measures of the angles in a polygonal cycloid generated by rolling an n-gon on a line?

15.6. Show that the length of the polygonal cardioid generated by an n-gon is $8R \cos^2(\pi/n) + 8r$, where r and R are the inradius and circumradius of the n-gon.

15.7. Show that every chord in a cardioid that passes through the cusp has length equal to four times the radius of the generating circle.

CHAPTER **16**

Star Polygons

Above the cloud with its shadow is the star with its light.

Pythagoras

Even a small star shines in the darkness.

Finnish proverb

Star polygons have long fascinated humans. Since antiquity they have been used as mythical and religious symbols, and they figure prominently in Judaism, Islam, and Christianity. Star-shaped objects are found in nature as well as in manufactured items and as symbols of achievement. In Figure 16.1 we see a starfish, some star-shaped bicycle sprockets, and a star on the Walk of Fame on Hollywood Boulevard.

Figure 16.1.

In addition to the stars in Figure 16.1, there are star-shaped flowers, stars on flags, and star logos of organizations and businesses. Stars are commonly used to indicate quality (e.g., five star restaurants), and are common shapes for cookies, pastries, and pasta. Stars with points are often used to represent the astronomical bodies called stars in the sky.

Star pentagons and hexagons were studied by the ancient Greek geometers, and the pentagram became the symbol of the Pythagorean school. The first persons to consider star polygons in general were Thomas Bradwardine (1290–1349), an English mathematician and briefly Archbishop of

Canterbury, and later Regiomontanus (1436–1476) and Charles de Bouelles (1470–1533). In conjunction with his study of polyhedra, Johannes Kepler (1571–1630) discussed *stellated polygons* in his book *Harmonices Mundi* (The Harmony of the World), published in 1619.

We begin with a discussion of the geometry of star polygons, and then deal with pentagrams, hexagrams, and star octagons. We conclude with a discussion of "magic stars" in recreational mathematics.

16.1   The geometry of star polygons

Ordinary (convex or concave) polygons are simple in the sense that their sides do not intersect. When they do we obtain *star polygons*. A regular star polygon, or *polygram*, is denoted by $\{p/q\}$ and is constructed as follows. Connect every qth vertex of a regular polygon with p sides, where $1 < q < p/2$. In Figure 16.2 we have, from left to right, the regular star polygons $\{5/2\}$, $\{6/2\}$, $\{7/2\}$, $\{7/3\}$, $\{8/2\}$, and $\{8/3\}$. Many regular star polygons have names: $\{5/2\}$ is a *pentagram*, $\{6/2\}$ is a *hexagram*, *Star of David*, or *Solomon's seal*, $\{8/2\}$ is the *star of Lakshmi*, and $\{8/3\}$ the *octagram*.

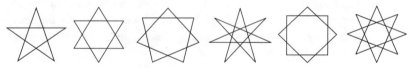

Figure 16.2.

A star polygon includes the internal line segments as well as the outline. For example, the pentagram is the first symbol in Figure 16.2, not the star ☆ (called a *mullet* in heraldry), which is a concave decagon.

For completeness we let $\{p/1\}$, or simply $\{p\}$, denote the regular convex polygon with p sides. When p and q are relatively prime, $\{p/q\}$ is *unicursal*, meaning it can be drawn on paper without lifting the pencil or retracing any segments. When p and q have a greatest common divisor $d > 1$, then $\{p/q\}$ consists of d copies of $\{(p/d)/(q/d)\}$, and the term *star figure* is often used. For example, the hexagram $\{6/2\}$ consists of two equilateral triangles $\{3\}$, and the star of Lakshmi $\{8/2\}$ consists of two squares $\{4\}$.

If we extend each side of an ordinary polygon in each direction until it intersects another side, we obtain a *stellated polygon*, and the process of creating such star polygons is called *stellation*. All the star polygons in Figure 16.2 are stellated polygons. In some cases each side is extend until the first intersection with the extension of another side, and in others until the second intersection with the extension of another side, as with $\{7/3\}$ and $\{8/3\}$.

Stars and law enforcement badges

Stars with five, six, seven, and occasionally eight points are common badge designs for law enforcement officers in the United States. The association between stars and law enforcement may date back to the use of stars in English and Scottish heraldry, where they were often associated with the use of spurs. In Figure 16.3 we see badges based on star polygons $\{p/2\}$ for $p = 5, 6, 7$, and 8.

Figure 16.3.

Thomas Bradwardine discovered the formula $(p - 2q)180°$ for the sum of the vertex angles of $\{p/q\}$. Each of the p vertex angles measures $[1 - 2(q/p)]180°$, and hence the sum is $(p - 2q)180°$. Thus the sum of the vertex angles of the pentagram is $180°$, of the hexagram and octagram $360°$, and the star of Lakshmi $720°$.

For general star pentagons the sum of the vertex angles is also $180°$, as can be seen in Figure 16.4, where we have [Nakhli, 1986]

$$\alpha + (\beta + \delta) + (\gamma + \varepsilon) = \alpha + \theta + \phi = 180°.$$

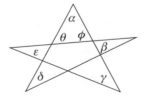

Figure 16.4.

We now show that Bradwardine's formula holds for a general (non-regular) star polygon $[p/q]$ (which is constructed from a general convex polygon analogously to $\{p/q\}$). Assume p and q are relatively prime, such as for $[7/3]$ in Figure 16.5a. Imagine an ant traversing the star polygon, starting at A, walking to B, turning through the marked exterior angle, walking to C, again turn through the marked exterior angle, etc., as in

Figure 16.5b. When the ant returns to A, it faces in the same direction as when it started the walk. The total turning is three complete revolutions, since visiting every third vertex requires three circuits of the vertices to visit every vertex.

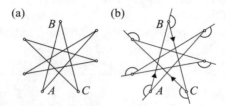

Figure 16.5.

Hence the sum of the exterior angles is $3 \cdot 360°$, each interior angle is $180°$ minus an exterior angle, so the sum of the seven interior angles is $7 \cdot 180° - 3 \cdot 360° = (7 - 2 \cdot 3)180°$, When we replace $[7/3]$ by $[p/q]$ we obtain $(p - 2q)180°$ for the sum [de Villiers, 1999].

When p and q have a greatest common divisor $d > 1$, then $[p/q]$ consists of d copies of $[(p/d)/(q/d)]$ star polygons, and the angle sum is the same as before:

$$d \cdot [(p/d) - 2(q/d)]180° = (p - 2q)180°.$$

As with the angles, there are nice relationships for the lengths of line segments in a general pentagram. In Figure 16.6a we see pentagram $ABCDE$ where we have labeled the sides of the small triangle at vertex A as a_1, a_2, and a_3, and similarly for the other vertices.

With the sides so labeled, we have

$$a_1 b_1 c_1 d_1 e_1 = a_2 b_2 c_2 d_2 e_2 \qquad (16.1)$$

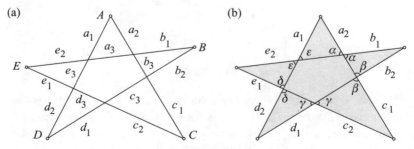

Figure 16.6.

and

$$\frac{(a_3 + b_1)}{(a_2 + b_3)} \cdot \frac{(b_3 + c_1)}{(b_2 + c_3)} \cdot \frac{(c_3 + d_1)}{(c_2 + d_3)} \cdot \frac{(d_3 + e_1)}{(d_2 + e_3)} \cdot \frac{(e_3 + a_1)}{(e_2 + a_3)} = 1 \quad (16.2)$$

To prove (16.1) we apply the law of sines to each of the gray triangles in Figure 16.6b to obtain $a_1/a_2 = \sin \alpha / \sin \varepsilon$, $b_1/b_2 = \sin \beta / \sin \alpha$, $c_1/c_2 = \sin \gamma / \sin \beta$, $d_1/d_2 = \sin \delta / \sin \gamma$, and $e_1/e_2 = \sin \varepsilon / \sin \delta$. Multiplying the equations yields $a_1 b_1 c_1 d_1 e_1 / a_2 b_2 c_2 d_2 e_2 = 1$ [Lee, 1998]. Equation (16.2) is proved similarly, see Challenge 16.1. Equation (16.1) also holds for general $[p/2]$ star polygons with a product of p terms on each side.

Star polygons in Gaudí's columns

The architect Antoni Gaudí (1852–1926), after careful studies in geometry, designed the columns in the Church of the Sagrada Familia in Barcelona, Spain. To avoid classical Greek and Solomonic columns and inspired by the helical shape of living trees, Gaudí used star polygons with 6, 8, 10, and 12 sides. He discovered [Bonet, 2000] that columns could have a star polygonal base with a unidirectional twist to the right, and intersect a second column with a twist to the left.

Figure 16.7.

The intersection of the two columns results in a column whose base is a star polygon but whose cross-sections at different heights are star polygons with more and more sides until they become close to circular. To avoid the angular vertices of the star polygons Gaudí rounded them with parabolas so that the cross-sections are smooth curves.

16.2   The pentagram

The pentagram, or regular star pentagon, is intimately connected with the *golden ratio* ϕ, the positive root $(\sqrt{5}+1)/2$ of the quadratic equation $x^2 - x - 1 = 0$. This follows from the fact that the diagonal of the regular pentagon with side 1 is equal to ϕ, which we now show using Figure 16.8.

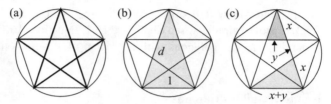

Figure 16.8.

Inscribe a regular pentagon with side 1 and diagonal d and a pentagram in a circle, as shown in Figures 16.8ab. Since the three angles at each vertex of the pentagon are equal (each subtends an arc equal to one-fifth of the circle), the shaded isosceles triangles in Figures 16.8bc are similar. Hence

$$d = \frac{d}{1} = \frac{x}{y} = \frac{x+y}{x} = 1 + \frac{y}{x} = 1 + \frac{1}{d},$$

so that $d^2 = d + 1$ and consequently $d = \phi$.

Since each of the vertex angles of the pentagram is 36° (because the five sum to 180°) we can use the pentagram from Figure 16.8a to find trigonometric functions of multiples of 18° [Bradie, 2002]. Draw a pair of parallel lines as shown in Figure 16.9 to create the two shaded right triangles, each with hypotenuse of length 1.

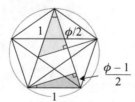

Figure 16.9.

In the light gray triangle, we have

$$\cos 36° = \sin 54° = \frac{\phi}{2} = \frac{\sqrt{5}+1}{4},$$

and in the dark gray triangle,

$$\sin 18° = \cos 72° = \frac{\phi - 1}{2} = \frac{\sqrt{5} - 1}{4}.$$

A pentagon with an inscribed pentagram yields a simple proof that the golden ratio ϕ is irrational. Assume ϕ is rational, and write $\phi = m/n$, where m and n are positive integers and the fraction is in lowest terms. Draw the pentagram inside a pentagon with side n and diagonal m as illustrated in Figure 16.10a. As shown in Figure 16.8c, the shaded triangle in Figure 16.10b is similar to the shaded triangle in Figure 16.10a, and hence $\phi = n/(m - n)$, a contradiction since $1 < \phi < 2$ implies $n < m$ and $m - n < n$.

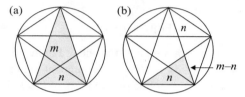

Figure 16.10.

To construct a paper model of a regular dodecahedron (a solid with twelve pentagonal faces, on the left in Figure 16.11) we usually use a *net*, the diagram shown in the center of the figure. Each half of the net can be drawn from a regular pentagon by inscribing a pentagram and then drawing a second pentagram in the center, extending the lines to mark the shaded triangular regions not a part of the net, as illustrated on the right in the figure [Bishop, 1962].

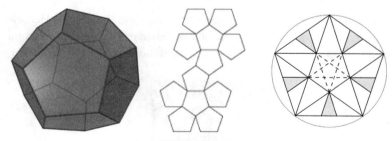

Figure 16.11.

Once we have a model of the dodecahedron, we can use pentagrams again to create a model of the *small stellated dodecahedron*, a star polyhedron with twelve pentagramic faces. The small stellated dodecahedron in Figure 16.12a is from a mosaic in St. Marks' Cathedral in Venice, created in about 1430 by Paolo Uccello (1397-1475). The drawing of the small stellated dodecahedron in Figure 16.12b is by Johannes Kepler (1571–1630).

(a)   (b)

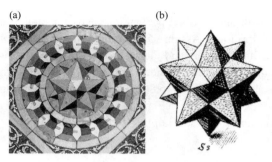

Figure 16.12.

If we take a paper pentagram such as the one in Figure 16.7a and fold the five triangular points upwards we form a pentagonal pyramid, and twelve such pyramids can be pasted onto the twelve faces of the dodecahedron. We need only verify that the triangular faces of the pyramids lie in the planes of the pentagonal faces of the dodecahedron to conclude that the model has pentagramic faces. The dihedral angle (the angle between the planes of adjacent faces) of the regular dodecahedron is $\arccos(-1/\sqrt{5}) \approx 116.565°$, and the dihedral angle between the base and a triangular face of the pyramid is $\arccos(1/\sqrt{5}) \approx 63.435°$, and hence the faces of the model are pentagrams.

16.3   The Star of David

The Leningrad Codex, the oldest surviving complete *Tanakh*, or Hebrew Bible, dates from 1008 or 1009. The Star of David or hexagram appears on the cover, as illustrated in Figure 16.13. Its Hebrew name, *Magen David*,

Figure 16.13.

means Shield of David. Today, the Star of David is the universally recognized symbol of the Jewish people, and appears on the flag of the state of Israel.

The Star of David hexagram is often viewed as the union of two equilateral triangles, one pointing upwards and the other downwards. The points where the sides of the triangles cross trisect the sides of the triangles forming another regular hexagon, illustrated in Figure 16.14.

Figure 16.14.

Overlaying a grid of smaller equilateral triangles enables us to make area comparisons. For example, the area of the hexagram is two-thirds the area of the original hexagon, and the area of the inner hexagon is one-half the area of the hexagram.

We conclude with an intriguing appearance of a hexagram when the sides of a triangle are trisected and the vertices joined to the trisection points, as shown in Figure 16.15a.

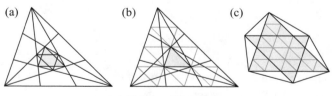

Figure 16.15.

The trisection lines form a non-regular hexagon in the center of the triangle and its diagonals yield a non-regular hexagram. The surprise is that the area of the hexagram is precisely $7/100$ the area of the original triangle.

We first find the area of a gray triangle inscribed in the hexagon, as illustrated in Figure 16.15b. Overlaying a grid of smaller triangles similar to the original shows that the area of the gray triangle is $1/16$ the area of the original triangle. In Figure 16.15c we see an enlarged version of the hexagram from Figure 16.15a, overlaid with another grid of similar triangles. Counting small triangles shows that the area of the hexagram is $28/25$ times the area of the gray triangle from Figure 16.15b, and hence the area of the hexagram is $(28/25) \cdot (1/16) = 7/100$ the area of the original triangle.

Chinese checkers and star numbers

Chinese checkers is a board game usually played with marbles on a wooden board with holes arranged in a hexagram pattern, as illustrated in Figure 16.16a. The standard board has $S_5 = 121$ holes, where S_n is the nth *star number*. For $n \geq 2$, $S_n = 12T_{n-1} + 1 = 6n(n-1) + 1$ where T_n is the nth triangular number (see Section 14.2), as can be seen in Figure 16.16b. The second star number $S_2 = 13$ appears in the Great Seal of the United States on the one-dollar bill, as shown in Figure 16.16c.

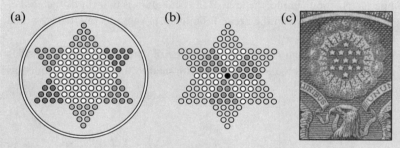

Figure 16.16.

In spite of its name, Chinese checkers did not originate in China and is unrelated to ordinary checkers. The game was invented in Germany in the 1890s and first produced in the United States in 1928. The name "Chinese checkers" was chosen for marketing purposes.

Another star hexagon is the so-called *unicursal hexagram*, illustrated in Figure 16.17. It is not regular, since its edges do not all have the same length. The sum of its vertex angles is $240°$.

Figure 16.17.

16.4 The star of Lakshmi and the octagram

Two regular star polygons can be constructed from the octagon—{8/2}, the *star of Lakshmi*, and {8/3}, the *octagram*. See Figure 16.18.

(a) 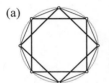 (b)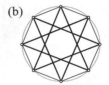

Figure 16.18.

In Indian philosophy, the star of Lakshmi symbolizes *Ashtalakshmi*, the eight forms of wealth of the Hindu goddess of fortune and prosperity Lakshmi, and plays a symbolic role in Hindu and Islamic designs.

Star octagons have appeared in arts and crafts all over the world for centuries. On the left in Figure 16.19 we have an {8/2} star from the Patio del Cuarto Dorado in the Alhambra built in the 14th century. Next we see an {8/3} star within an {8/2} star, surrounded by eight more {8/2} stars, from the Real Alcázar in Seville, also built in the 14th century. In the third part of the figure we see a commonly found hex sign signifying abundance and goodwill from the eastern Pennsylvania Amish community of the mid-19th century. On the right we have a traditional quilt block pattern known as the Lone Star, popular with American quilters.

Figure 16.19.

Two interesting constants, the *Córdoba proportion* and the *silver ratio*, are found in the star octagons. To illustrate, we need expressions for the side lengths of the {8/2} and {8/3} stars, i.e., the lengths of the diagonals of a regular octagon. They are easily computed using right triangles, as shown in Figure 16.20. If we let the side of the octagon be 1 and d_1, d_2, and d_3 the diagonals shown then it follows that $d_1 = \sqrt{2 + \sqrt{2}}$, $d_2 = 1 + \sqrt{2}$, and $d_3 = \sqrt{4 + 2\sqrt{2}}$.

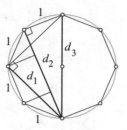

Figure 16.20.

Consequently, the circumradius of the star of Lakshmi (and the octagram as well) is $R = d_3/2 = \sqrt{4 + 2\sqrt{2}}/2 = 1/\sqrt{2 - \sqrt{2}} \approx 1.3065\cdots$. This was called the *Córdoba proportion* by the Spanish architect Rafael de la Hoz, who discovered that this ratio appears in many Arab and Moorish buildings, such as the Mezquita mosque in Córdoba [de la Hoz, 1995].

The side length $d_2 = 1 + \sqrt{2}$ of the octagram has been called the *silver ratio* because of properties it shares with the golden ratio ϕ (see Section 16.2). A common symbol for the silver ratio is δ_S. Both ϕ and δ_S appear as side lengths of star polygons, and both have simple continued fraction expressions:

$$\phi = 1 + \cfrac{1}{1 + \cfrac{1}{1 + \frac{1}{1+\cdots}}} \quad \text{and} \quad \delta_s = 2 + \cfrac{1}{2 + \cfrac{1}{2 + \frac{1}{2+\cdots}}}.$$

Since $\phi - 1 = 1/\phi$, when a square is removed from a *golden rectangle* (one whose dimensions are $\phi \times 1$), a similar rectangle remains, as seen in Figure 16.21a. Similarly, $\delta_S - 2 = 1/\delta_S$ implies that when two squares are removed from a *silver rectangle* (one whose dimensions are $\delta_S \times 1$), a similar rectangle remains, as seen in Figure 16.21b.

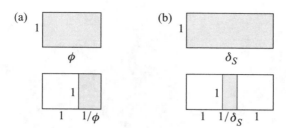

Figure 16.21.

Non-regular octagrams appear in the study of centroids of quadrilaterals. If we place four equal masses at the vertices of the convex quadrilateral $ABCD$ in Figure 16.22a, the *centroid* or *center of gravity O* is the center of the *Varignon parallelogram PQRS*, whose vertices are the midpoints of the sides of $ABCD$.

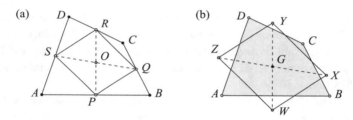

Figure 16.22.

However, if the convex quadrilateral $ABCD$ is made of uniform material, then the center of gravity is the center G of the *Wittenbauer parallelogram XYZW* in Figure 16.22b, which is determined by the trisection points of the sides of $ABCD$. For a proof see [Foss, 1959]. Because their sides are parallel to the diagonals of $ABCD$, $PQRS$ and $XYZW$ are parallelograms as claimed.

The octagram as a compass rose

The octagram colored as shown in Figure 16.23a is called a *compass rose*, and is used on maps, nautical charts, and magnetic compasses to indicate direction. The octagram as compass rose can be found on the marker for the "zero point" of France, a spot near the Cathedral of Notre Dame from which all highway distances in France are measured. See Figure 16.23b.

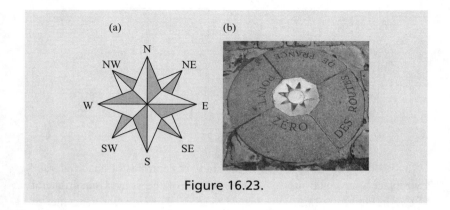

Figure 16.23.

16.5   Star polygons in recreational mathematics

There is a variety of recreational mathematic problems where star polygons appear. They have been popularized by Martin Gardner [Gardner, 1975]. Magic stars are similar to their better-known cousins, magic squares. The object is to place numbers in cells so that the number sums are the same in every direction.

For example, the template for the *magic pentagram* is shown in Figure 16.24a, where we are asked to place the numbers 1 through 10 in the circles so that the sum of the four numbers in the five lines is the same. The sum is called the *magic constant* for the magic pentagram.

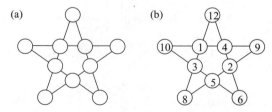

Figure 16.24.

The magic constant is easily found. Since the numbers 1 through 10 sum to 55 and each number appears in two lines, the sum of the five line sums is $2 \cdot 55 = 110$. Since each of the line sums is the same, the magic constant (if the magic pentagram exists) is $110/5 = 22$.

However, magic pentagrams do not exist. The following simple argument is due to Ian Richards [Gardner, 1975].

The two lines containing the number 1 must contain six other numbers summing to 21. Since $9 + 8 + 7 + 6 + 5 + 4 = 39$, 1 and 10 must lie on the same line, call it A. Let B be the other line through 1 and C the other line through 10. If $A = \{1, 10, 4, 7\}$, then it is impossible to find sets of four numbers for B and C. So there are three possibilities:

A	B	C
1, 10, 2, 9	1, 6, 7, 8	10, 5, 4, 3
1, 10, 3, 8	1, 5, 7, 9	10, 6, 4, 2
1, 10, 5, 6	1, 4, 8, 9	10, 7, 3, 2

In none of them do B and C have a number in common, and hence magic pentagrams do not exist.

If we relax the restriction that we must use the numbers from 1 through 10, but still require that all numbers be positive and distinct, then defective magic pentagram do exist. Figure 16.24b shows one constructed from $\{1, 2, 3, 4, 5, 6, 8, 9, 10, 12\}$ with magic constant 24.

Figure 16.25 shows that magic hexagrams, heptagrams, and octagrams exist, with magic constants 26, 30, and 34, respectively.

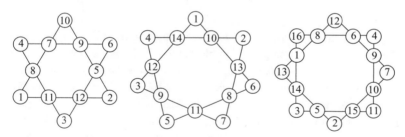

Figure 16.25.

It can be shown that there are 80 different magic hexagrams, 72 different heptagrams, and 112 different octagrams.

A different type of magic hexagon is constructed by entering the numbers 1 through 12 in the twelve triangular cells of the hexagram in Figure 16.26a so that the five numbers in each of the indicated six rows have the same sum.

The solution, described in [Bolt et al., 1991; Gardner, 2000], culminates in a computer search and consists of just two configurations: those in Figures 16.26b and 16.26c with magic constants 33 and 32.

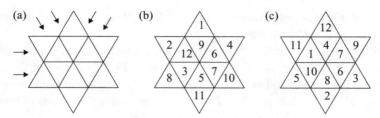

Figure 16.26.

Another recreational mathematics problem is the *tree planting problem*, or *orchard planting problem*, specifying the number of trees, number of rows, and number of trees in each row to be planted. For example, the problem of planting 10 trees in 5 rows with 4 trees per row can be solved with the pentagram as well as with other configurations, as illustrated in Figure 16.27.

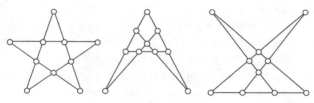

Figure 16.27.

Let $r(n, k)$ be the number of rows given n trees and k trees per row. The problem of finding the largest possible value of $r(n, k)$ is unsolved. The configurations in Figure 16.27 illustrate $r(10, 4) = 5$. The hexagram shows how to plant 12 trees in 6 rows with 4 trees per row (see Figure 16.25), but that is not maximal since $r(12, 4) = 7$ for the configuration shown in Figure 16.28a. However, configurations based on star polygons often illustrate the maximal number of rows, e.g., $r(16, 4) = 15$ as illustrated in Figure 16.28b [Dudeney, 1907].

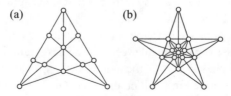

Figure 16.28.

16.6 Challenges

16.1. For the pentagram $ABCDE$ illustrated in Figure 16.6a, prove (16.2).

16.2. Suppose $ABCDE$ is a pentagon (not necessarily regular) with its inscribed pentagram, as shown in Figure 16.29, with the property that each of the five triangles ABC, BCD, CDE, DEA, and EAB has area 1. Show that the area of the pentagon is $\phi + 2$, where ϕ is the golden ratio.

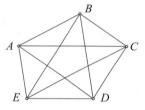

Figure 16.29.

16.3. By joining each vertex of a square to the midpoints of the nonadjacent sides we can create a non-regular star octagon, as shown in Figure 16.30a.

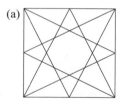 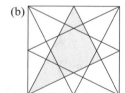 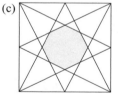

Figure 16.30.

Show that (a) the shaded triangle in Figure 16.30b is a right triangle with sides in the ratio 3:4:5 and (b) the area of the shaded octagon in Figure 16.30c has 1/6 the area of the original square.

16.4. The defective magic pentagram in Figure 16.24b has the numbers 1, 2, 3, 4, and 5 in its central pentagon. Is it possible to construct a defective magic pentagram with the same ten numbers but with 1, 2, 3, 4, and 5 in the vertices of the pentagram?

16.5. Prove that in a magic hexagram, the sum of the three numbers at the corners of each large triangle must be the same.

16.6. In 1821 John Jackson published the following problem in verse [Burr, 1981]:

> Your aid I want, nine trees to plant
> In rows just half a score,
> And let there be in each row three,
> Solve this: I ask no more.

(Hint: "Half a score" is ten.)

16.7. Suppose 49 trees are growing at the points of a 7-by-7 grid, as shown in Figure 16.31. (a) Is it possible to harvest 29 of the trees, and leave the remaining 20 in 18 rows with 4 trees in each row? [R. Bracho López, 2000] (b) Is it possible to harvest 39 of the trees and leave the remaining 10 in 5 rows with 4 trees in each row? (Hint: see Figures 16.30a and 16.28 to get started.)

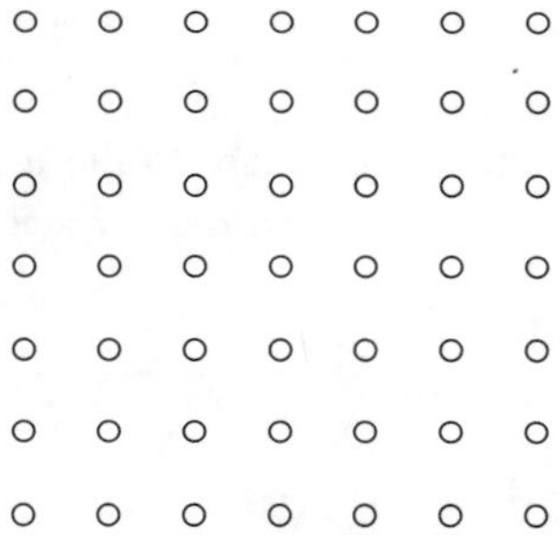

```
o   o   o   o   o   o   o

o   o   o   o   o   o   o

o   o   o   o   o   o   o

o   o   o   o   o   o   o

o   o   o   o   o   o   o

o   o   o   o   o   o   o

o   o   o   o   o   o   o
```

Figure 16.31.

16.8. Within a circle inscribe a regular decagon and its $\{10/3\}$ star, as shown in Figure 16.32. Prove that the side length of the $\{10/3\}$ star is equal to the side length of the decagon plus the radius of the circle.

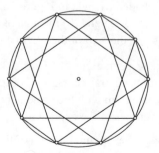

Figure 16.32.

16.9. Here are two puzzles just for fun from Sam Loyd's *Cyclopedia* [Loyd, 1914].

    (a) In "The Lost Star" puzzle, Loyd asks us to find a perfect five-pointed star (☆) in Figure 16.32a. Find it.

    (b) In "The New Star" puzzle, Loyd asks us "to show how and where to place another star of the first magnitude" in Figure 16.32b. By "first magnitude" Loyd means larger than all the others, and not touching any of the others. Solve the puzzle.

(a)   (b)

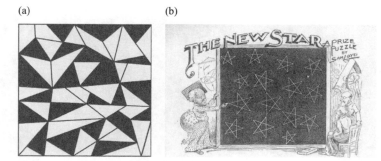

Figure 16.33.

16.10. In Figure 16.32 we see the unicursal hexagram from Figure 16.17 inscribed in a rectangle. If the indicated edges are perpendicular, what is the ratio of the dimensions of the rectangle?

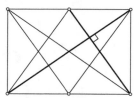

Figure 16.34.

CHAPTER 17

Self-similar Figures

> *Great fleas have little fleas upon their backs to bite 'em,*
> *And little fleas have lesser fleas, and so* ad infinitum.
> *And the great fleas themselves, in turn, have greater fleas to*
> *go on,*
> *While these again have greater still, and greater still, and so*
> *on.*
>
> Augustus De Morgan
> *A Budget of Paradoxes* (1872)

An object is called *self-similar* when it is similar to a proper subset of itself, such as the partitioned equilateral triangle in the icon for this chapter. Self-similar objects also look the same when magnified or reduced in size. Approximately self-similar natural objects include the romanesco broccoli, the shell of the chambered nautilus, and the flower of the tromsø palm seen in Figure 17.1.

Figure 17.1.

In mathematics we can produce self-similar figures through *iteration*, repeating a process to subdivide a figure into similar parts or grow it by adjoining a region to produce a figure similar to the original. We will consider a variety of geometrical figures in which one part of the figure is a scaled version of the entire figure.

221

Iterating the bride's chair

In Chapter 1 we encountered the bride's chair icon, a right triangle surrounded by three squares. To iterate it, use each leg square as the hypotenuse square of a similar brides chair. After four iterations, we have the object on the left in Figure 17.2.

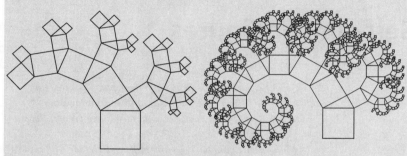

Figure 17.2.

Continuing yields a lovely fractal known as the *Pythagoras tree*, as shown on the right in Figure 17.2. For an introduction to the mathematics of fractals, see [Mandlebrot, 1977].

17.1   Geometric series

In Section 6.1 we employed similar right triangles to illustrate the familiar formula for the sum of a geometric series with a positive first term a and common ratio $r < 1$: $a + ar + ar^2 + \cdots = a/(1 - r)$. When $a = r = 1/n$, $n \geq 2$, we can we can use self-similar figures to illustrate

$$\frac{1}{n} + \left(\frac{1}{n}\right)^2 + \left(\frac{1}{n}\right)^3 + \cdots = \frac{1/n}{1 - (1/n)} = \frac{1}{n - 1}. \qquad (17.1)$$

For $n = 2$ we have $1/2 + 1/4 + 1/8 + \cdots = 1$, which we illustrate with the partition of the unit square in Figure 17.3.

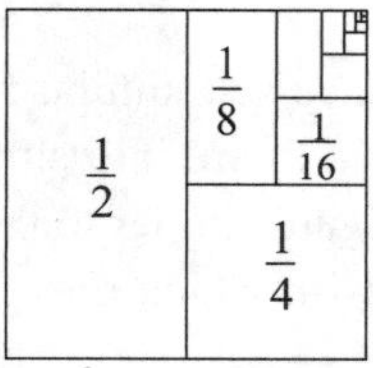

$$
\begin{array}{|c|c|c|}
\hline
 & \dfrac{1}{8} & \\
\dfrac{1}{2} & & \dfrac{1}{16} \\
 & \dfrac{1}{4} & \\
\hline
\end{array}
$$

Figure 17.3.

For $n = 3$ we have $1/3 + 1/9 + 1/27 + \cdots = 1/2$, which we illustrate with a rectangle with area 1 whose sides are in the ratio $\sqrt{3}$ to 1. The gray region is congruent to the white region and hence has area 1/2 [Mabry, 2001].

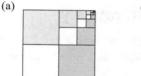

Figure 17.4.

For $n = 4$ we illustrate $1/4 + 1/16 + 1/64 + \cdots = 1/3$ in two ways in Figure 17.5 [Ajose, 1994; Mabry, 1999].

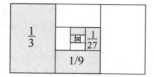

Figure 17.5.

For $n = 5$ consider a right triangle with legs $1/\sqrt{5}$ and $2/\sqrt{5}$, and hence with hypotenuse 1 and area $1/5$. Four of them and a square with area $1/5$ form a unit square, as shown in Figure 17.6a.

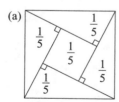
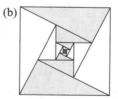

Figure 17.6.

If we partition the square with area $1/5$ into four right triangles and a square each with area $1/25$ and continue in this fashion, we obtain $1/5 + 1/25 + 1/125 + \cdots = 1/4$, as shown in Figure 17.6b.

In general, (17.1) can be illustrated with a regular $(n-1)$-gon partitioned into n congruent trapezoids and a similar regular $(n-1)$-gon whose edge length is $1/\sqrt{n}$ the edge length of the original [Tanton, 2008]. In

Figure 17.7a we see an octagon (an 8-gon) where the smaller octagon in the center has an edge length 1/3 that of the original.

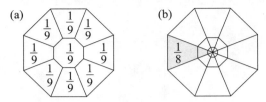

Figure 17.7.

When we repeat the process, the series of trapezoids in the gray region of Figure 17.7b illustrates $1/9 + 1/81 + 1/729 + \cdots = 1/8$. For illustrations of geometric series using right triangles, see Challenge 17.5.

17.2 Growing figures iteratively

Some finite sums can be illustrated by growing a figure by iteration. While the result is only approximately self-similar, it does enable us to evaluate the sum.

For example, in Figure 17.8 we see arrays with 1, 3, 9, 27, and 81 dots, obtained iteratively by replacing each dot in the kth array with a triangle of three dots to obtain the $(k + 1)$st [Sher, 1997].

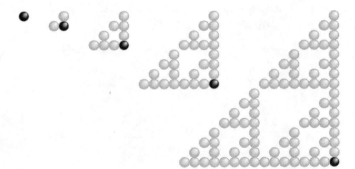

Figure 17.8.

If we count the 81 dots in a different way, those above and below the diagonal line in Figure 17.9, we see that for $n = 3$, $1 + 2(1 + 3 + 3^2 + \cdots + 3^n) = 3^{n+1}$.

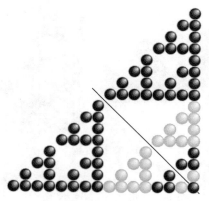

Figure 17.9.

Hence we have

$$1 + 3 + 3^2 + \cdots + 3^n = \frac{3^{n+1} - 1}{2}.$$

To illustrate the corresponding formula for powers of 2, we fold a piece of string in half n times, as shown in Figure 17.10 for $n = 4$. The right side of the picture is, in a sense, a scaled version of the portion in the upper half of the left side of the picture.

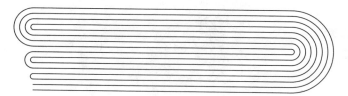

Figure 17.10.

On the right we see one group of 2^n strings and on the left the results of preceding folds, so $1 + 1 + 2 + 2^2 + \cdots 2^{n-1} = 2^n$, or [Tanton, 2009],

$$1 + 2 + 2^2 + \cdots 2^{n-1} = 2^n - 1.$$

In Figure 17.11 we have square arrays of 1, 4, 16, 64, and 256 balls, each constructed from the preceding one by adjoining three copies of itself [Sher, 1997].

Figure 17.11.

The right-most array is square with $1 + 1 + 2 + 2^2 + \cdots 2^n = 2^{n+1}$ balls on a side for $n = 3$, but counting the balls by the way in which they are shaded yields

$$1 + 3(1 + 4 + 4^2 + \cdots + 4^n) = (2^{n+1})^2 = 4^{n+1}$$

and hence

$$1 + 4 + 4^2 + \cdots + 4^n = \frac{4^{n+1} - 1}{3}.$$

Similarly, we can grow triangular arrays of balls to illustrate identities for triangular numbers $T_k = 1 + 2 + \cdots + k$ when k is a power of 2.

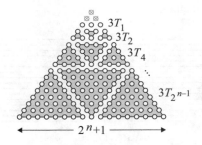

Figure 17.12.

In Figure 17.12 we see a triangular array of T_{2^n+1} balls illustrating

$$3(T_1 + T_2 + T_4 + \cdots + T_{2^{n-1}}) + 3 = T_{2^n+1},$$

or [Nelsen, 2005],

$$T_1 + T_2 + T_4 + \cdots + T_{2^{n-1}} = \frac{1}{3}T_{2^n+1} - 1. \qquad (17.2)$$

17.3 Folding paper in half twelve times

It is folk wisdom that it is impossible to fold a sheet of paper, no matter how large or how thin, in half more than seven times. A standard sheet of copier paper measures (in inches) 8.5×11 and is about .04 inch thick. If we fold it in half seven times, then the result is a pile of paper less than $3/4$ of a square inch in area and about one-half inch thick. It is impossible to fold this pile in half.

What if we replace the alternate-direction folding of a rectangular sheet of paper with same-direction folding of a very long but narrow piece of paper, such as a ribbon or a roll of toilet paper? Then the folding yields a pile of paper that, when viewed from the side, resembles Figure 17.10. This is what Britney Gallivan, then a high school student in California, did in 2002 when challenged to fold a sheet of paper in half twelve times. First Britney calculated how long a strip of paper she needed [Gallivan, 2002].

She considered the loss of paper due to folding, that is, the paper consumed by the folds in the semicircular parts of the diagram in Figure 17.10. Let t denote the thickness of the paper, n the number of times the paper is folded in half, and L_n the loss of paper length due to folding. The total length L of paper required satisfies $L \geq L_{12}$. For $n = 4$ Figure 17.10 shows that
$L_4 = \pi t + (\pi t + 2\pi t) + (\pi t + 2\pi t + 3\pi t + 4\pi t) + (\pi t + 2\pi t + \cdots + 8\pi t)$,
and in general we have

$$L_n = \pi t (T_1 + T_2 + T_4 + \cdots + T_{2^{n-1}}).$$

From (17.2),

$$L_n = \frac{\pi t}{3}(T_{2^n+1} - 3) = \frac{\pi t}{6}[(2^n + 1)(2^n + 2) - 6] = \frac{\pi t}{6}(2^n + 4)(2^n - 1).$$

Single ply toilet paper is approximately 2 mils ($t = .002$ inch) thick, which yields $L_{12} \approx 1465$ feet. If we want to have, say, 6 inches of paper in each of the 2^{12} layers between the folds, we need to add about another 2048 feet, for a total length of approximately 3513 feet. Britney used a roll of paper 4000 feet long to fold a piece of paper twelve times. In Figure 17.13 we see Britney and the paper prior to the twelfth fold.

Figure 17.13.

17.4   The *spira mirabilis*

The *logarithmic spiral* (or *equiangular* or *growth* spiral) is a spiral that appears in nature. It was studied by René Descartes (1596–1650) and later by Jacob Bernoulli (1654–1705) who gave it the name *spira mirabilis* ("miraculous spiral"). Its equation in polar coordinates is $r = ae^{b\theta}$ where a and b are non-zero constants. When $b > 0$ it has the appearance of the graph in Figure 17.14a, while when $b < 0$ it winds clockwise.

Figure 17.14.

Among the approximate appearances of the spira mirabilis in nature are the shells of the chambered nautilus in Figure 17.1, cloud patterns in a region of low pressure in Figure 17.14b, and the arms of a spiral galaxy such as the Whirlpool galaxy in the constellation Canes Venatici in Figure 17.14c.

The spira mirabilis is a self-similar object. If we scale the graph of $r_1 = ae^{b\theta}$ by the number $e^{2\pi b}$ to obtain $r_2 = e^{2\pi b} \cdot r_1 = ae^{b(\theta+2\pi)}$, the graph remains the same, as illustrated in Figure 17.15.

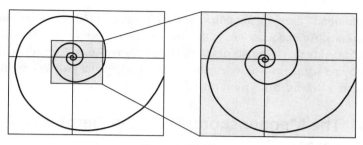

Figure 17.15.

If (s, ϕ) is on the graph of $r_1 = ae^{b\theta}$ then so is $(s, \phi + 2\pi)$ (since (s, ϕ) and $(s, \phi + 2\pi)$ are coordinates of the same point in the polar plane), and hence (s, ϕ) is also on the graph of $r_2 == ae^{b(\theta + 2\pi)}$. Similarly any point on the graph of $r_2 = e^{2\pi b} \cdot r_1$ is also on the graph of $r_1 = ae^{b\theta}$.

The self-similarity property of the spira mirabilis yields several interesting consequences. The points of intersection of the spiral with a ray $\theta = \theta_0$ partition the ray into line segments whose lengths are in a geometric progression. Furthermore, the tangent lines to the spiral at the points of intersection with the ray all make the same angle with the spiral, which justifies the name *equiangular spiral*.

When θ is increased by $\pi/2$, the distance of a point on the spira mirabilis from the pole in multiplied by $e^{b\pi/2}$. If this number is the golden ratio ϕ, the spiral has the polar equation $r = a\phi^{2\theta/\pi}$ and can be inscribed in a *golden rectangle*, a rectangle where the ratio of the sides is ϕ. Such a spiral is called a *golden spiral*. It could be confused with the *Fibonacci spiral*, constructed by inscribing quarter circles in a set of squares whose sides are Fibonacci numbers. In Figure 17.16a we see a portion of a Fibonnaci spiral inscribed in a rectangle whose sides are in the ratio of 21 to 34, which approximates a golden rectangle. Fibonacci spirals also appear on the 1987 Swiss postage stamp and the 10 litas Lithuanian gold coin minted in 2007, as shown in Figures 17.16b and 17.16c.

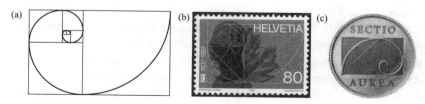

Figure 17.16.

In addition to the spira mirabilis and the Fibonacci spiral, there are many other spirals. An important one is the *Archimedean spiral* ($r = a\theta$), which has the property that its points of intersection with a ray $\theta = \theta_0$ partition the ray into line segments whose lengths are in an arithmetic progression. See [Alsina and Nelsen, 2010] for further details.

17.5 The Menger sponge and the Sierpiński carpet

The *Menger sponge* is a curious object first described by Karl Menger (1902–1985) in 1926. It is constructed from a cube with side length 1 iteratively. Divide the cube into 27 ($3 \times 3 \times 3$) smaller cubes, and remove the center cube and the six cubes at the middle of each face. Now repeat the process with the remaining 20 cubes. The Menger sponge is the limit after infinitely many iterations. In Figure 17.18a we see Menger and the portion of the cube remaining after four iterations.

Figure 17.17.

With each iteration the volume of the remaining portion of the cube decreases, while the surface area increases. After n iterations the remaining volume of the cube is $V_n = (20/27)^n$ and the surface area is $A_n = 2(20/9)^n + 4(8/9)^n$, thus one of the surprising properties of the Menger sponge is that is has zero volume but infinite surface area!

Each face of the Menger sponge is called a *Sierpiński carpet*, after the Polish mathematician Waclaw Sierpiński (1882–1969) who first described it in 1916 (see Figure 17.18 for pictures of Sierpiński and the carpet after four iterations). The Sierpiński carpet has zero area, but the perimeter of the holes is infinite (see Challenge 17.4). For more properties of the Menger sponge and Sierpiński carpet, see [Mandelbrot, 1977].

Figure 17.18.

17.6 Challenges

17.1. What series and their sums are illustrated by the partitions of the square and triangle (assuming each has area 1) in Figure 17.19?

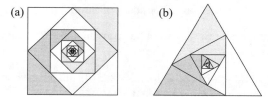

(a)   (b)

Figure 17.19.

17.2. In Figure 17.20 we see a placement of five queens on a 5-by-5 chessboard so that no queen is attacking another. Use iteration to show how to construct a maximal placement of non-attacking queens on an infinite chessboard.

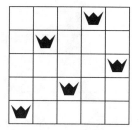

Figure 17.20.

17.3. One of the prettiest fractals is *Sierpiński's triangle*, also known as the Sierpiński sieve or gasket. It was first described by Sierpiński in 1915. To construct one, begin with an equilateral triangle and delete the central one-fourth, then delete the central one-fourth of the remaining smaller equilateral triangles, and continue. The first four steps are shown in Figure 17.21. The Sierpiński triangle is the limit after infinitely many iterations.

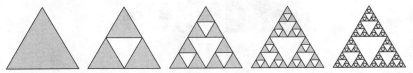

Figure 17.21.

The Sierpiński triangle is a self-similar object, consisting of three copies of itself, scaled by a factor of 1/2. Show that the Sierpiński triangle has zero area but a boundary of infinite length.

17.4. Show that the Sierpiński carpet has zero area, but the perimeter of the holes is infinite.

17.5. Create illustrations of (17.1) for $n = 2, 3, 4$, and 5 based on the similar right triangles in Figures 5.14 and 5.15.

17.6. A *golden rectangle* (a $1 \times \phi$ rectangle, where $\phi = (1 + \sqrt{5})/2$ is the golden ratio) has the property that if we cut off a square from it (as in Figure 17.22b), the new rectangle is similar to the original. The process can be continued indefinitely (see Figure 17.22c). Conclude that

$$1 + \frac{1}{\phi^2} + \frac{1}{\phi^4} + \frac{1}{\phi^6} + \cdots = \phi.$$

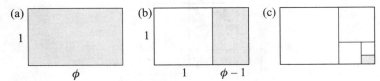

Figure 17.22.

CHAPTER **18**

Tatami

The bright harvest moon
Upon the tatami mats
Shadows of the pines.
Takarai Kikaku (1661–1707)

Tatami mats are a type of flooring used in traditional Japanese homes. In the Muromachi period (1332–1573), tatami had a core made of rice straw (today it might be wood chips or styrofoam) covered in woven rush. See Figure 18.1a. Tatami flooring is cool in the summer, warm in the winter, and well suited to the humid months in Japan. The size of a tatami mat varies by region, but they are rectangular, with one dimension twice the other, usually about 1 meter by 2 meters, and about 5-1/2 to 6 cm thick. Sizes of rooms are often measured by number of tatami mats, and in Figure 18.1b we see tatami mat layouts for 12, 8, and 4-1/2 (4×6 m, 4×4 m, and 3×3 m) mat rooms. In the last case, a square half mat is used, and we show two arrangements. Arrangements of the mats vary, but an auspicious design employs T-shaped junctions of mats rather than +-shaped junctions.

(a)   (b)

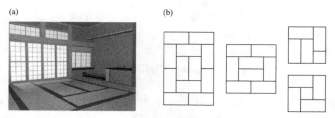

Figure 18.1.

Our icon is based on the symmetric arrangement of tatami mats in a 4-1/2 mat room in the lower right corner of Figure 18.1b. It consists of four a-by-b rectangles ($b > a > 0$) similarly arranged in a square $a + b$ units on each side. We will also consider other arrangements of mats that completely cover the floor of a room without any of the mats overlapping.

The Catalan architect Antoni Gaudí, discussed in Chapter 16, used designs similar to our tatami icon in some of his buildings. In Figure 18.2 we see door panels and a window screen from the Casa Vicens in Barcelona.

Figure 18.2.

18.1   The Pythagorean theorem—Bhāskara's proof

Let a and b (with $b > a > 0$) denote the legs and c the hypotenuse of a right triangle, and construct the tatami icon with four $a \times b$ rectangles. Partition each rectangle into two right triangles as shown in Figure 18.3a.

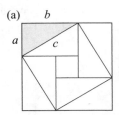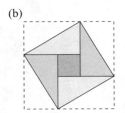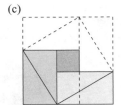

Figure 18.3.

In Figures 18.3b and 18.3c we discard four triangles in two different ways, leaving shaded regions with the same area. The area of the shaded square in Figure 18.3b is c^2, while the shaded region in Figure 18.3c is the union of two squares with areas a^2 and b^2. Thus $c^2 = a^2 + b^2$.

Figures 18.3b and 18.3c without the dashed lines constitute an elegant one-word proof of the Pythagorean theorem attributed to the Indian mathematician and astronomer Bhāskara (1114–1185). The one word: *Behold!*

We conclude with another property of right triangles that can be demonstrated easily with the tatami icon. Arrange four right triangles so that their hypotenuses form a square, as shown in Figure 18.4a. Then the vertices of the four right angles lie on a line segment, as shown in Figure 18.4b by adding the dashed lines to complete the tatami icon. Since the line segment is a diagonal of the icon, it bisects the square and the right angles at the far left and right in Figure 18.4a (cf. Figure 2.3) [DeTemple and Harold, 1996]].

(a)   (b)

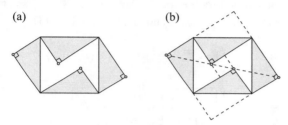

Figure 18.4.

18.2 Tatami mats and Fibonacci numbers

Suppose we have a collection of 1-by-2 tatami mats, and we wish to cover a 2-by-n hallway for some $n \geq 1$. Each mat can be placed so that its longer side is parallel to or perpendicular to the direction of the hallway. Let t_n denote the number of arrangements of tatami mats in a 2-by-n hallway. For example, $t_5 = 8$—see Figure 18.5 for a picture of a 2-by-5 hallway and the eight ways to arrange the tatami mats.

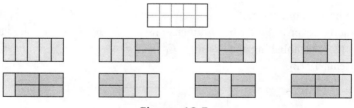

Figure 18.5.

We have $t_1 = 1, t_2 = 2, t_3 = 3, t_4 = 5, t_5 = 8, t_6 = 13$, and so on. It is easy to see that for any $n \geq 3, t_n = t_{n-1} + t_{n-2}$: to cover the hallway, start at one end with a tatami mat perpendicular to the direction of the hallway and complete the covering in t_{n-1} ways, or start the covering with a pair of tatami mats parallel to the direction of the hallway and complete the covering in t_{n-2} ways.

So if we define $t_0 = 1$ (there is only one way to cover a hallway with zero area—use no tatami mats), then for all $n \geq 0$, $t_n = F_{n+1}$, the $(n+1)$ st Fibonacci number. See *Proofs That Really Count* [Benjamin and Quinn, 2003] for a collection of lovely proofs of Fibonacci identities using similar methods.

The square tatami icon can also be used to illustrate Fibonacci identities. If we let the dimensions of the rectangle in the tatami icon in Figure 18.6a be F_{n-1} and F_n, then the side of the small central square is F_{n-2} and the side of the large enclosing square is F_{n+1}, and thus [Bicknell and Hoggatt, 1972].

$$F_{n+1}^2 = 4F_n F_{n-1} + F_{n-2}^2.$$

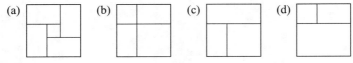

Figure 18.6.

We can construct other illustrations for Fibonacci identities by covering floors with mats whose dimensions are Fibonacci numbers. If the dimensions of the large squares in Figures 18.6bcd are $F_{n+1} = F_n + F_{n-1}$, then we have [Ollerton, 2008].

$$
\begin{aligned}
F_{n+1}^2 &= F_n^2 + 2F_n F_{n-1} + F_{n-1}^2, \\
&= F_n^2 + F_n F_{n-1} + F_{n+1} F_{n-1}, \\
&= F_{n-1}^2 + F_n F_{n-1} + F_{n+1} F_n.
\end{aligned}
$$

If we have a collection of n square mats with areas $F_1^2, F_2^2, \ldots, F_n^2$, they will exactly cover a floor whose dimensions are F_n-by-F_{n+1}, as shown in Figure 18.7, establishing the identity $F_1^2 + F_2^2 + \cdots + F_n^2 = F_n F_{n+1}$ [Brousseau, 1972].

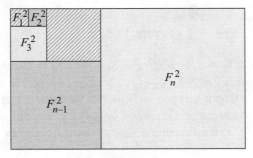

Figure 18.7.

18.3 Tatami mats and representations of squares

In Figure 18.8a we see a square 9-by-9 room covered with 40 1-by-2 tatami mats and one 1-by-1 half mat in the center. If we compute the area covered by the tatami mats by concentric rings of mats, we have

$$9^2 = 1 + 4 \cdot 2 + 8 \cdot 2 + 12 \cdot 2 + 16 \cdot 2$$
$$= 1 + 8(1 + 2 + 3 + 4).$$

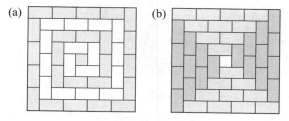

Figure 18.8.

In general, for any odd number $2n + 1$, we have

$$(2n + 1)^2 = 1 + 8(1 + 2 + 3 + \cdots + n) = 1 + 8T_n. \qquad (18.1)$$

See Figure 18.8b, and note that the area of each shaded region is twice a triangular number [Wakin, 1987]. As a consequence we have another derivation of the formula for the nth triangular number T_n:

$$T_n = 1 + 2 + 3 + \cdots + n = \frac{(2n + 1)^2 - 1}{8} = \frac{n(n + 1)}{2}.$$

Similarly, Figure 18.9a illustrates a 10-by-10 room covered by 50 1-by-2 tatami mats, and computing the area by concentric rings of mats yields

$$10^2 = 2 \cdot 2 + 2 \cdot 6 + 2 \cdot 10 + 2 \cdot 14 + 2 \cdot 18.$$

Dividing by 4 yields $5^2 = 1 + 3 + 5 + 7 + 9$ as illustrated in Figure 18.9b.

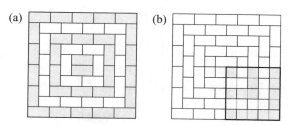

Figure 18.9.

In general, $1 + 3 + 5 + \cdots + (2n - 1) = n^2$.

One way to represent k^3 in the plane is to use k copies of a k-by-k square mat. Consider the mats covering the floor in the 30-by-30 room in Figure 18.10 [Lushbaugh, 1965; Cupillari, 1989].

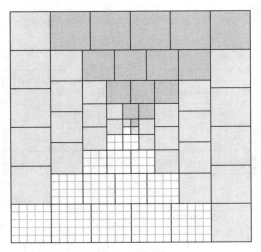

Figure 18.10.

The sum of the areas of the white mats is $1^3 + 2^3 + 3^3 + 4^3 + 5^3$, which equals one-fourth of the area of the room, i.e., $30^2/4 = [(5 \cdot 6)/2]^2$. In general, we have a formula for the sum of the first n cubes: $1^3 + 2^3 + \cdots + n^3 = [n(n + 1)/2]^2$.

18.4   Tatami inequalities

In Sections 4.5 and 13.4 we encountered inequalities for the arithmetic mean $(a + b)/2$, the geometric mean $\sqrt{ab}$, and the harmonic mean $2ab/(a + b)$ of two positive numbers a and b:

$$\frac{a + b}{2} \geq \sqrt{ab} \geq \frac{2ab}{a + b}, \tag{18.2}$$

with equality if and only if $a = b$. We illustrate this with the tatami icon.

In Figure 18.11 we see the icon with four tatami mats whose sides are in the ratio a to b, but scaled to fit square rooms of different sizes. In Figure 18.11a, the side of the room is the arithmetic mean $(a + b)/2$, and each mat has area $ab/4$. Since the four mats may not completely cover the floor, we have $((a + b)/2)^2 \geq 4(ab/4) = ab$, which, on taking square roots, yields the first inequality in (18.2) [Schattschneider, 1986].

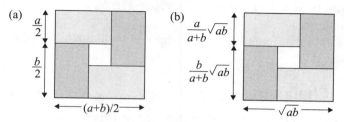

Figure 18.11.

In Figure 18.11b, the mats are scaled so that the side of the room is the geometric mean $\sqrt{ab}$, and each mat has area $(ab/(a + b))^2$. The four mats may not completely cover the floor, so $ab \geq 4(ab/(a + b))^2$, which, on taking square roots, yields the second inequality in (18.2). In each case the mats cover the floor if and only if they are squares, i.e., $a = b$.

18.5 Generalized tatami mats

Although we have seen proofs of the addition formula for the sine in Chapters 2 and 3, we present another using tatami mats, similar to our proof of the Pythagorean theorem. Suppose we have a pair of tatami mats measuring $\sin \alpha$ by $\cos \alpha$ and another pair measuring $\sin \beta$ by $\cos \beta$ for positive first quadrant angles α and β. The diagonals of each mat are 1, and they can be arranged in a $(\sin \alpha + \sin \beta)$ by $(\cos \alpha + \cos \beta)$room (along with a small mat to fill the center) as shown in Figure 18.12a.

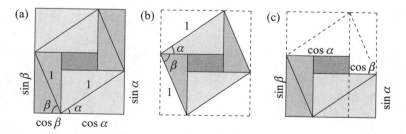

Figure 18.12.

Cut the four larger mats along a diagonal and discard four of the resulting triangular mats in two different ways, as shown in Figures 18.12b and

18.12c. The two remaining shaded regions have equal area, $\sin(\alpha + \beta)$ for the parallelogram in Figure 18.12b, and $\sin\alpha \cos\beta + \cos\alpha \sin\beta$ for the sum of the two adjoining rectangles in Figure 18.12c, which establishes the addition formula.

The subtraction formula for the cosine can be illustrated similarly; see Challenge 18.3.

18.6   Challenges

18.1. Iteration of the 4-1/2 tatami mat icon in Figure 18.1 yields Figure 18.13. If the area of the entire icon is 1, what infinite series (and its sum) is illustrated?

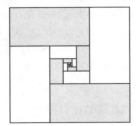

Figure 18.13.

18.2. (a) Cover the floor of the L-shaped room in Figure 18.14 in two different ways with tatami mats whose dimensions are Fibonacci numbers to show that $F_{n+1}^2 + F_n^2 = F_{n-1}F_{n+1} + F_n F_{n+2}$.

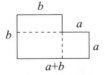

Figure 18.14.

(b) Use the result in (a) to prove *Cassini's identity*: for all $n \geq 2$, $F_{n-1}F_{n+1} - F_n^2 = (-1)^n$.

18.3. Illustrate $\cos(\alpha - \beta) = \cos\alpha \cos\beta + \sin\alpha \sin\beta$ using generalized tatami mats similar to those in Section 18.5.

18.4. Another puzzle from Sam Loyd's *Cyclopedia* [Loyd, 1914] is "The Juggler" (see Figure 18.15). Loyd asks: "Cut one of the triangles in half and then fit the six pieces into a perfect square." Solve the puzzle. (Hints: The right triangles are congruent with one leg twice the other, and by "in half" Loyd means "into two pieces.")

Figure 18.15.

18.5. Let p and q be two positive numbers whose sum is 1. Prove that

$$\text{(a) } \frac{1}{p} + \frac{1}{q} \geq 4 \text{ and (b) } \left(p + \frac{1}{p} \right)^2 + \left(q + \frac{1}{q} \right)^2 \geq \frac{25}{2}.$$

(Hint: use a tatami figure for (a) and a figure with overlapping squares for (b).)

18.6. It is easy to show that $T_8 = 6^2$ and $T_{288} = 204^2$. Prove that there are infinitely many numbers that are simultaneously triangular and square. (Hint: Consider T_{8T_n} and (8.1).)

18.7. Bhāskara's proof of the Pythagorean theorem in Section 18.1 may have motivated the following puzzle from Henry Ernest Dudeney's classic book *Amusements in Mathematics* (1917): "Take a strip of paper, measuring five inches by one inch, and, by cutting it into five pieces, the parts fit together and form a square, as shown in Figure 18.16. Now, it is quite an interesting puzzle to discover how we can do this in only four pieces." Solve the puzzle.

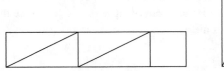

Figure 18.16.

18.8. Since the overlapping squares icon is similar to the tatami icon, it is not surprising that it also illustrates Fibonacci identities. Use overlapping squares and rectangles to illustrate the following: (a) $F_{n+1}^2 = 2F_{n+1}F_{n-1} + F_n^2 - F_{n-1}^2$, (b) $F_{n+1}^2 = 2F_n^2 + 2F_{n-1}^2 - F_{n-2}^2$, and (c) $F_{n+1}^2 = 4F_n^2 - 4F_{n-1}F_{n-2} - 3F_{n-2}^2$.

CHAPTER **19**

The Rectangular Hyperbola

> *What mathematician has ever pondered over an hyperbola,*
> *mangling the unfortunate curve with lines of intersection*
> *here and there, in his efforts to prove some property...?*
>
> Lewis Carroll
> *The Dynamics of a Parti-cle* (1865)

Hyperbolas, along with circles, ellipses, and parabolas, belong to the family of plane curves known as conic sections. Menaechmus, Euclid, and Aristaeus studied conic sections, and Apollonius of Perga (circa 240-190 BCE) wrote *Conics*, an eight-volume work that presented the modern form of characterizing these curves as intersections of cones with planes.

Hyperbolas appear in nature and in man-made objects. From left to right in Figure 19.1 we see the hyperbolic path taken by some single-apparition comets, three of the six hyperbolic arcs at the tip of a sharpened hexagonal pencil, and the hyperbolic shape of the light beam of a flashlight held parallel to a wall.

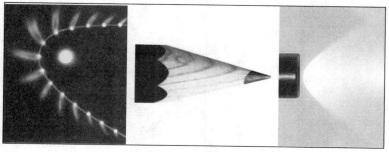

Figure 19.1.

Striking examples of the use of the hyperbola in architecture appear in the design of the columns of the Catedral Metropolitana Nossa Senhora Aparecida in Brasília, and in the profile of the McDonnell Planetarium in St. Louis, as illustrated in Figure 19.2.

Figure 19.2.

A *rectangular hyperbola* (or *equilateral* or *right hyperbola*) is one whose asymptotes are perpendicular. We will explore the role played by rectangular hyperbolas in the mathematics of logarithms, inequalities, and hyperbolic functions.

Hyperbolas and bells

The first book addressing the problem of finding the ideal shape for a bell was *Harmonicorum, libri XII* (1648), a beautiful study by Marin Mersenne (1588–1648) searching for the functional description of the profile of a bell.

Figure 19.3.

Mersenne's description of bells with a hyperbolic shape has influenced the design of bells for centuries, and may be considered as a starting point for the parameterization of curves.

19.1 One curve, many definitions

There are a variety of ways to define a hyperbola, including (a) as an intersection of a cone (with two nappes) and a plane, (b) the focus-directrix definition, and (c) the two-foci definition, which we encountered in Section 11.2. They are equivalent. See [Eves, 1983; Alsina and Nelsen, 2010] for an elegant proof, due to the Belgian mathematicians Adolphe Quetelet (1796–1874) and Germinal-Pierre Dandelin (1794–1847), that the conic section and focus-directrix definitions are equivalent.

Rectangular hyperbolas are often defined by equations, such as $xy = k \neq 0$ (with the lines $xy = 0$, i.e., the x- and y-axes, as asymptotes) or $x^2 - y^2 = c \neq 0$ (with the lines $x^2 - y^2 = 0$, i.e. $y = x$ and $y = -x$ as asymptotes). It is simple but tedious to show that they satisfy the two-foci definition. Unlike hyperbolas in general, there is another nice definition for the rectangular hyperbola $xy = k^2$: the first-quadrant branch of the hyperbola is the locus of the fourth vertex of all rectangles with a fixed area k^2 with one vertex at the origin and two vertices on the positive x- and y-axes. See Figure 19.4.

Figure 19.4.

19.2 The rectangular hyperbola and its tangent lines

We begin with a simple optimization problem. What line through (a, b) in the first quadrant cuts off the triangle in the first quadrant with minimum area? We claim it is the line L with equation $(x/a) + (y/b) = 2$ as illustrated in Figure 19.5a. Observe that (a, b) is the midpoint of the portion of L lying in the first quadrant.

To prove that L minimizes the area of the triangle, take another line through (a, b), such as the dashed line in Figure 19.5b. It creates a triangle

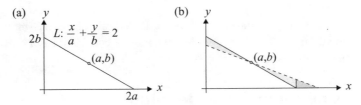

Figure 19.5.

whose area is larger by the area of the small dark gray triangle, since the two light gray triangles are congruent and have the same area.

For the rectangular hyperbola $xy = ab$ through (a, b), as shown in Figure 19.6a, we claim that the line L in Figure 19.5a is its tangent line at (a, b). To prove this, we show that (a, b) is the only point common to the hyperbola and the line (as the portion of L in the first quadrant has finite length). Our proof is from [Lange, 1976].

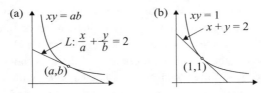

Figure 19.6.

Assume (a', b'), different from (a, b), also lies on L and the hyperbola. Hence $(a'/a) + (b'/b) = 2$ and $a'b' = ab$, or equivalently,

$$\frac{(a'/a) + (b'/b)}{2} = 1 \quad \text{and} \quad \sqrt{\frac{a'}{a} \cdot \frac{b'}{b}} = 1.$$

The arithmetic and geometric means of the numbers a'/a and b'/b are the same, so they must be equal. Thus $ab' = a'b$ which, when combined with $a'b' = ab$, yields $a = a'$ and $b = b'$, a contradiction.

In Figure 19.6b we see the special case $(a, b) = (1, 1)$ where the hyperbola $y = 1/x$ lies on or above the tangent line $y = 2 - x$, so that for all positive x

$$x + \frac{1}{x} \geq 2, \tag{19.1}$$

i.e., *the sum of a positive number and its reciprocal is at least* 2. The inequality follows immediately from the AM-GM inequality but (19.1) also implies the AM-GM inequality, so the two statements are equivalent. To see this, let $x = \sqrt{a}/\sqrt{b}$ in (19.1) for any two positive numbers a and b and clear fractions to yield $a + b \geq 2\sqrt{ab}$, which is equivalent to the AM-GM inequality.

The rectangular hyperbola can also be used to construct the geometric mean of two positive numbers. See Challenge 19.1.

19.3   Inequalities for natural logarithms

In virtually every modern calculus text (and presumably in most calculus classrooms) the natural logarithm $\ln b$ of a positive number b is defined as the definite integral

$$\ln b = \int_1^b \frac{1}{x}\, dx.$$

Interpreting the integral as area, if $0 < a < b$ then the area under the graph of $xy = 1$ over the interval $[a, b]$ is $\ln(b/a) = \ln b - \ln a = \int_a^b (1/x)\, dx$.

Grégoire de St. Vincent

The Belgian mathematician Grégoire de St. Vincent (1584–1667) published *Opus Geometricum* in 1647, where he applied Archimedes' method for finding the area of a parabolic segment to regions bounded by hyperbolas. He showed that the area under the graph of $xy = k$ over the intervals $[a, b]$ and $[c, d]$ are the same if $a/b = c/d$ (a, b, c, d, k positive). Hence areas under a rectangular hyperbola behave like logarithms. He accomplished this using properties of the conic sections, without using coordinate geometry or calculus [Burn, 2000].

Approximations to areas of regions bounded by an arc of $xy = 1$ lead to interesting inequalities for natural logarithms. For example, if we bound the region under the graph of $xy = 1$ over the interval $[a, b]$ with inscribed and circumscribed rectangles and compute areas, as shown in Figure 19.7, we obtain *Napier's inequality*:

$$\text{if } 0 < a < b, \quad \text{then} \quad \frac{1}{b} < \frac{\ln b - \ln a}{b - a} < \frac{1}{a}. \tag{19.2}$$

Comparing areas of the rectangles and the region under the hyperbola yields

$$\frac{1}{b}(b - a) < \int_a^b \frac{1}{x}\, dx < \frac{1}{a}(b - a)$$

from which Napier's inequality follows.

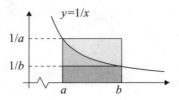

Figure 19.7.

Napier's inequality can be used to derive a familiar limit expression for the number e. Let $a = n$, $b = n + 1$, apply some elementary algebra, and take the limit as $n \to \infty$ using the squeeze theorem [Schaumberger, 1972]:

$$\frac{1}{n + 1} < \ln \frac{n + 1}{n} < \frac{1}{n}$$

$$\ln e^{n/(n+1)} = \frac{n}{n + 1} < \ln \left(1 + \frac{1}{n}\right)^n < 1 = \ln e$$

$$e^{n/(n+1)} < \left(1 + \frac{1}{n}\right)^n < e$$

$$\lim_{n \to \infty} \left(1 + \frac{1}{n}\right)^n = e.$$

Taking reciprocals in (19.2) yields

$$\text{if } 0 < a < b, \quad \text{then} \quad a < \frac{b - a}{\ln b - \ln a} < b,$$

that is, $(b - a)/(\ln b - \ln a)$ is a mean of a and b, naturally called the *logarithmic mean* of a and b. To compare the logarithmic mean to the arithmetic and geometric means, we use better approximations to the area under the graph of $xy = 1$ over the interval $[a, b]$. See Figure 19.8.

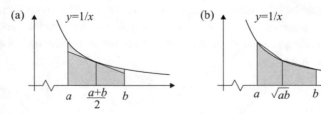

Figure 19.8.

In Figure 19.8a the area of the inscribed trapezoid is $2(b - a)/(a + b)$ and consequently

$$\frac{2(b - a)}{a + b} < \ln b - \ln a \quad \text{or} \quad \frac{b - a}{\ln b - \ln a} < \frac{a + b}{2}.$$

In Figure 19.8b we circumscribe two trapezoids, one with base $[a, \sqrt{ab}]$ and one with base $[\sqrt{ab}, b]$. The area of each trapezoid simplifies to $(b - a)/2\sqrt{ab}$ and hence

$$\frac{b - a}{\sqrt{ab}} > \ln b - \ln a \quad \text{or} \quad \sqrt{ab} < \frac{b - a}{\ln b - \ln a}.$$

We obtain the same inequalities when $0 < b < a$, and consequently the logarithmic mean lies between the geometric and arithmetic means, i.e., for positive a and b,

$$\sqrt{ab} \leq \frac{b - a}{\ln b - \ln a} \leq \frac{a + b}{2}. \tag{19.3}$$

with equality if and only if $a = b$.

As an application of (19.3), let $\{a, b\} = \{1, x\}$ for x positive, $x \neq 1$. Then

$$\sqrt{x} < \frac{x - 1}{\ln x} < \frac{1 + x}{2} \quad \text{and hence} \quad \lim_{x \to 1} \frac{\ln x}{x - 1} = 1.$$

For another application, see Challenge 19.3.

19.4 The hyperbolic sine and cosine

In most calculus texts, the hyperbolic sine and cosine are defined in terms of the exponential functions as $\sinh \theta = (e^\theta - e^{-\theta})/2$ and $\cosh \theta = (e^\theta + e^{-\theta})/2$. Then certain identities are verified and the source of the name "hyperbolic" is revealed: the point $(\cosh \theta, \sinh \theta)$ lies on the right-hand branch of the rectangular hyperbola $x^2 - y^2 = 1$. But why choose those combinations of e^θ and $e^{-\theta}$?

The circular sine and cosine are usually defined as coordinates of points on the unit circle, that is, $(\cos \theta, \sin \theta)$ is the point on $x^2 + y^2 = 1$ for which the signed area of the shaded circular sector in Figure 19.9a is $\theta/2$.

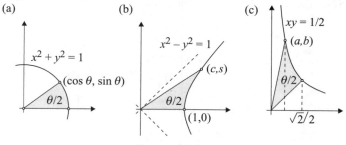

Figure 19.9.

Let (c, s) be the point on the right-hand branch of $x^2 - y^2 = 1$ for which the signed area of the shaded hyperbolic sector in Figure 19.9b is $\theta/2$. We now prove that $s = (e^\theta - e^{-\theta})/2 = \sinh\theta$ and $c = (e^\theta + e^{-\theta})/2 = \cosh\theta$. Since $c^2 - s^2 = 1$, we seek a second equation in c and s, and Figure 19.9b suggests expressing the area $\theta/2$ of the hyperbolic sector in terms of c and s.

To do so, rotate the hyperbola and the sector 45° counterclockwise, as illustrated in Figure 19.9c. The rotation takes the vertex $(1,0)$ of the hyperbola $x^2 - y^2 = 1$ to $(\sqrt{2}/2, \sqrt{2}/2)$, the point (c, s) to the point $(a, b) = (\sqrt{2}(c - s)/2, \sqrt{2}(c + s)/2$, and the equation of the hyperbola becomes $xy = 1/2$. The area of the shaded sector in Figure 19.9c is the same as the area under the hyperbola over the interval $(a, \sqrt{2}/2)$ (see Challenge 19.2), and hence

$$\frac{\theta}{2} = \int_a^{\sqrt{2}/2} \frac{1}{2x}\, dx = \frac{1}{2}[\ln(\sqrt{2}/2) - \ln a] = -\frac{1}{2}\ln(c - s),$$

so that $c - s = e^{-\theta}$. This and $c^2 - s^2 = 1$ yields the familiar expressions $s = (e^\theta - e^{-\theta})/2$ and $c = (e^\theta + e^{-\theta})/2$ for the hyperbolic sine and cosine.

19.5   The series of reciprocals of triangular numbers

The graph of a rectangular hyperbola can be used to illustrate the series

$$\frac{1}{1} + \frac{1}{3} + \frac{1}{6} + \cdots + \frac{1}{T_n} + \cdots = 2$$

where $T_n = 1 + 2 + 3 + \cdots + n$ is the nth triangular number. In Figure 19.10 we have used the graph of $xy = 2$ and the fact that $T_n = n(n + 1)/2$ to construct gray rectangles whose areas are the terms $\{1, 1/3, 1/6, \cdots\}$ of the series. Since the rectangles all have the same base they can be assembled, as

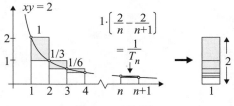

Figure 19.10.

shown on the right in the figure, into a larger rectangle with area 2, the sum of the series. Other telescoping series can be illustrated in a similar fashion. For example, see [Plaza, 2010].

19.6 Challenges

19.1. Let A and B be two points on the first-quadrant branch of $xy = 1$, let M be the midpoint of AB, and let P be the point of intersection of OM with the hyperbola, as shown in Figure 19.11. Prove that the coordinates of P are the geometric means of the corresponding coordinates of A and B.

Figure 19.11.

19.2. Let A and B be two points on the same branch of a rectangular hyperbola. Prove that the area of the hyperbolic sector bounded by the arc between A and B, as illustrated in Figure 19.12a, is the same as the area under the hyperbolic arc between A and B, as illustrated in Figure 19.12b.

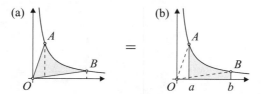

Figure 19.12.

19.3. Show that (a) for $x > -1$, $x \neq 0$, $x/(1 + x) < \ln(1 + x) < x$ and (b) for $x > 0$, $2x/(2 + x) < \ln(1 + x) < x/\sqrt{1 + x}$.

19.4. Let $x > 0$. Prove that $\lim_{n \to \infty} n\left(x^{1/n} - 1\right) = \ln x$.

19.5. Show that for $-1 < x < 1$ but $x \neq 0$, $x < \ln\sqrt{(1 + x)/(1 - x)} < x/\sqrt{1 - x^2}$.

19.6. Prove that for a and b positive, (19.3) can be sharpened to

$$\sqrt{ab} \leq \sqrt[4]{ab}\,\frac{\sqrt{a} + \sqrt{b}}{2} \leq \frac{b - a}{\ln b - \ln a}$$

$$\leq \left(\frac{\sqrt{a} + \sqrt{b}}{2}\right)^{2} \leq \frac{a + b}{2}. \qquad (19.4)$$

CHAPTER **20**

Tiling

The mathematician's patterns, like the painter's or poet's, must be beautiful.

G. H. Hardy
A Mathematician's Apology

Among the most beautiful of the mathematician's patterns are tilings. A *plane tiling* or *tessellation* is an arrangement of closed shapes that completely covers the plane without overlapping and without leaving gaps. Beautiful plane tilings abound in man-made objects, such as simple patterns for quilts and floors, the intricate mosaics found in Moorish buildings such as the Alhambra in Granada, and the ingenious designs of the Dutch graphic artist Maurits Cornelis Escher (1898–1972).

Three naturally occurring tilings are shown in Figure 20.1—a honeycomb, the columns of basalt in Giant's Causeway in Northern Ireland, and mud drying in a lakebed.

Figure 20.1.

Mathematicians have classified and studied a great variety of tilings—regular and semi-regular tilings, periodic, aperiodic, Penrose tilings, etc. However, our aim is to use tilings for performing mathematical operations and discovering mathematical results. In most instances we need only a portion of the entire plane tiling, as illustrated in the icon for this chapter.

20.1   Lattice multiplication

One of the earliest examples of the use of a tiling to do mathematics is an algorithm known as *lattice multiplication*, some times called *gelosia multiplication* (the pattern resembles a window grating or lattice, *gelosia* in Italian). Lattice multiplication may have originated in India sometime before the twelfth century, as it appears in a commentary on the *Lilāvati* of Bhāskara. Leonardo of Pisa (circa 1170–1250), also known as Fibonacci, introduced the method to Europe is his *Liber Abaci* (Book of Calculation) about 1202. The method uses a lattice or grid of isosceles right triangles, as illustrated in our icon for this chapter.

We illustrate the lattice multiplication algorithm with the example 63579× 523. The numbers 63579 and 523 are written along the top and right side of the lattice, and the products of the digits entered in each square with the tens digit above the units digit. After summing along the diagonals, the answer 33251817 is read down the left side and across the bottom of the lattice.

Figure 20.2.

John Napier (1550–1617), a Scotsman best known for inventing logarithms, also invented an abacus-like calculation tool based on lattice multiplication. Illustrated in Figure 20.13, it is known as *Napier's bones* or *Napier's rods*.

Figure 20.3.

20.2   Tiling as a proof technique

Every quadrilateral—convex or concave—tiles the plane, as illustrated in Figure 20.4.

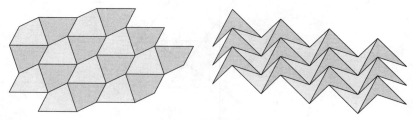

Figure 20.4.

Hence we have a simple procedure for finding the area of a quadrilateral: *The area of a quadrilateral Q is equal to one-half the area of a parallelogram P whose sides are parallel to and equal in length to the diagonals of Q.*

Figure 20.5 illustrates the result for convex quadrilaterals. Challenge 20.2 asks for a proof for concave quadrilaterals.

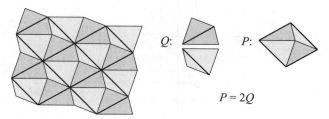

$P = 2Q$

Figure 20.5.

In Chapter 8 we encountered the Napoleon triangle—the equilateral triangle whose vertices were the centers of the three equilateral triangles erected on the sides of a general triangle. Is there anything similar for quadrilaterals?

The answer is yes for parallelograms and squares, which is illustrated in the following theorem: *the quadrilateral determined by the centers of the squares externally erected on the four sides of an arbitrary parallelogram is a square.*

See Figure 20.6 for a portion of the plane tiling generated by a parallelogram and the squares erected on its sides and an overlay created by connecting the centers of the squares [Flores, 1997].

If we let a and b denote the lengths of the sides of the parallelogram, P its area, and S the area of each overlay square, then we have the parallelogram analog of Napoleon's theorem: $2S = 2P + a^2 + b^2$.

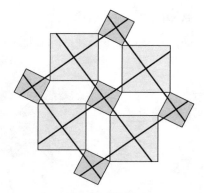

Figure 20.6.

20.3   Tiling a rectangle with rectangles

We say that a rectangle is tiled by rectangles if a collection of rectangles covers a rectangle with no overlaps or gaps, as illustrated in Figure 20.7.

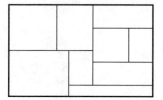

Figure 20.7.

In 1969 N. G. de Bruijn proved the remarkable result: *whenever a rectangle is tiled by rectangles each of which has at least one integer side, then the tiled rectangle has at least one integer side.* In the survey paper [Wagon, 1987] Stan Wagon presents fourteen different proofs of this result, including extensions, using a variety of techniques, from double integrals to graph theory. We present an elementary proof from [Konhauser et al., 1996], which Wagon attributes to R. Rochberg and S. K. Stein (independently).

Tile the first quadrant of the plane checkerboard fashion with gray and white squares, each measuring $1/2$ unit on a side, as shown in Figure 20.8. In Figure 20.8a we see that any rectangle with an integer side contains an equal amount of gray and white area. Consequently the tiled rectangle R contains an equal amount of gray and white area.

Place R in the first quadrant with one vertex at the origin, as shown in Figure 20.8b. If neither side of R is an integer, then R can be partitioned into

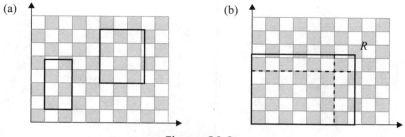

Figure 20.8.

four pieces (as shown by the dashed lines), three of which contain an equal amount of gray and white area, and one that doesn't. Hence at least one side of R must be an integer.

20.4   The Pythagorean theorem—infinitely many proofs

We began this book with Euclid's proof of the Pythagorean theorem and presented several more in subsequent chapters. We conclude by illustrating that there are infinitely many different proofs of the Pythagorean theorem.

In Chapter 1 we encountered several dissection proofs of the Pythagorean theorem, two of which are shown in Figure 20.9.

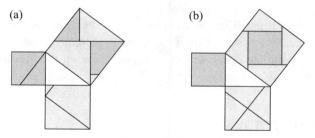

Figure 20.9.

How are such dissections proofs created? One way is to use a plane tiling based on squares of two different sizes, as shown in Figure 20.10. Such a tiling is often called the *Pythagorean tiling*, for reasons we shall now see.

In the Pythagorean tiling draw lines through the upper right hand corners of each of the smaller dark gray squares as shown in Figure 20.10a.

(a)   (b)

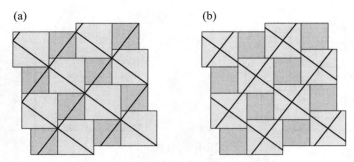

Figure 20.10.

This overlays on the tiling a grid of congruent transparent squares that we call the *hypotenuse grid*—note the right triangle formed in the lower left corner of the larger light gray squares. If we let a and b denote the legs and c hypotenuse of the right triangles, then the areas of the dark and light gray squares are a^2 and b^2, and the transparent squares formed by the hypotenuse grid illustrates the dissection proof that $c^2 = a^2 + b^2$ from Figure 20.9a.

Shift the hypotenuse grid so that the intersections of the lines lie at the centers of the larger light gray squares, as shown in Figure 20.10b. This yields the dissection proof of the Pythagorean theorem shown in Figure 20.9b.

Hence we can create as many different dissection proofs of the Pythagorean theorem as there are ways to overlay the hypotenuse grid on the Pythagorean tiling. Thus we have proved that there are infinitely many different dissection proofs of the Pythagorean theorem. In fact, there are uncountably infinitely many, and in each the square on the hypotenuse is dissected into nine or fewer pieces!

20.5   Challenges

20.1. Is it possible to tile the plane with figures bounded by circular arcs?

20.2. Prove that the area of a concave quadrilateral Q is equal to one-half the area of a parallelogram P whose sides are parallel to and equal in length to the diagonals of Q.

20.3. Floor tiles in the Salon de Carlos V in the Real Alcázar of Seville (Figure 20.11a) suggest another tiling proof of the Pythagorean theorem. Use the tiling with rectangles and squares and the overlay grid of squares (Figure 20.11b) to prove the theorem.

(a) (b)

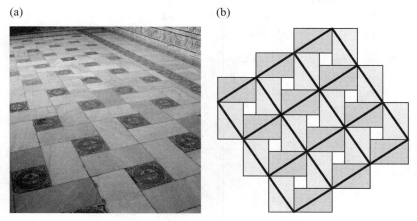

Figure 20.11.

20.4. If we inscribe a square in a semicircle and another square in a circle of the same radius (as shown in Figure 20.12), how do their areas compare? (Hint: Use a tiling consisting of unit squares and draw a circle of radius $\sqrt{5}$ centered at a point where four squares meet.)

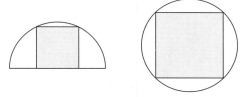

Figure 20.12.

20.5. Here is another puzzle from Henry Ernest Dudeney's *Amusements in Mathematics* (1917): "Farmer Wurzel owned the three square fields shown in the plan in Figure 20.13, containing respectively 18, 20, and 26 acres. In order to get a ring-fence round his property he bought the four intervening triangular fields. The puzzle is to discover what was then the whole area of his estate." (Hint: place

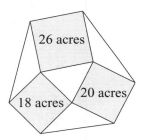

Figure 20.13.

the plan—a Vecten configuration—on a square measuring 144 acres, tiled with 1 acre squares.)

20.6. Use a tiling of the plane by $|a| \times |x|$ and $|b| \times |y|$ rectangles to prove the two-dimensional version of the Cauchy-Schwarz inequality: for real numbers a, b, x, y,

$$|ax + by| \leq \sqrt{a^2 + b^2}\sqrt{x^2 + y^2}.$$

Solutions to the Challenges

Many of the Challenges have multiple solutions. Here we give but one solution to each Challenge, and encourage readers to search for others.

Chapter 1

1.1. (a) Yes, but they are all similar to the 3-4-5 right triangle. If we set $b = a+d$ and $c = a+2d$, then $a^2+b^2 = c^2$ implies $a^2-2ad-3d^2 = 0$, so that $a = 3d$, $b = 4d$, and $c = 5d$. (b) Yes, and they are all similar to the right triangle with sides 1, $\sqrt{2}$, and $\sqrt{3}$. (c) If we set $b^2 = a^2r$ and $c^2 = a^2r^2$, then $r^2 = r + 1$ so that r must be the golden ratio $\phi = (1 + \sqrt{5})/2$.

1.2. Let x, y, z denote the sides of the outer squares, and let A, B, C denote the measures of the angles in the central triangle opposite sides a, b, c, respectively. For (a), the law of cosines yields $a^2 = b^2+c^2-2bc \cos A$ and $x^2 = b^2 + c^2 - 2bc \cos(180° - A) = b^2 + c^2 + 2bc \cos A$, hence $x^2 = 2b^2+2c^2-a^2$. Similarly $y^2 = 2a^2+2c^2-b^2$ and $z^2 = 2a^2 + 2b^2-c^2$, from which the result follows. In (b), we have $x^2+y^2+z^2 = 6c^2$ and $z^2 = c^2$, hence $x^2 + y^2 = 5z^2$.

1.3. The line segments $P_b P_c$ and AP_a are perpendicular and equal in length, as are the line segments $P_b P_c$ and AP_x, hence A is the midpoint of $P_a P_x$. The other two results are established similarly.

1.4. Label the vertices of the squares and triangles as shown in Figure S1.1, let the sides of the square $AEDP$ be a, and the sides of the square $CHIQ$ be b. Triangles ABE and BCH are congruent with legs a and b, and the corollary in the paragraph preceding Figure 1.7 tells us that triangles ABE and DEK have the same area, as do triangles BCH and KHI, and triangles KMN and DKI. Hence triangles ABE, DEK, KHI, and BCH all have the same area $(ab/2)$, and furthermore, $|PQ| = 2(a + b)$.

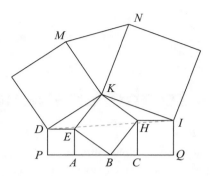

Figure S1.1.

Thus

$$\begin{aligned}
\text{area}(KMN) &= \text{area}(DKI) = \text{area}(PDKIQ) - \text{area}(PDIQ) \\
&= (a^2 + b^2 + 4 \cdot ab/2 + \text{area}(BEKH)) \\
&\quad -(1/2) \cdot 2(a + b) \cdot (a + b) \\
&= (a + b)^2 + \text{area}(BEKH) - (a + b)^2 \\
&= \text{area}(BEKH)
\end{aligned}$$

as claimed. ($\text{Area}(BEKH) = a^2 + b^2$.)

1.5. The areas of triangle ABH and quadrilateral $HIJC$ are equal. Let $a = |BC|$ and $b = |AC|$. Since triangles ACI and ADE are similar, $|CI|/b = a/(a + b)$ so that $|CI| = ab/(a + b)$ and $\text{area}(ACI) = ab^2/2(a + b)$. Similarly $\text{area}(BJC) = a^2b/2(a + b)$, and thus $\text{area}(ACI) + \text{area}(BJC) = ab/2 = \text{area}(ABC)$. Hence

$$\begin{aligned}
\text{area}(AJH) + 2\text{area}(HIJC) &+ \text{area}(BIH) \\
&= \text{area}(ACI) + \text{area}(BJC) \\
&= \text{area}(ABC) \\
&= \text{area}(AJH) + \text{area}(HIJC) \\
&\quad + \text{area}(BIH) + \text{area}(ABH),
\end{aligned}$$

and thus $\text{area}(HIJC) = \text{area}(ABH)$ [Konhauser et al., 1996].

1.6. Construct congruent parallelograms $ABED'$ and $ADFB'$, as shown in Figure S1.2 and observe that Q and S are the centers of $ABED'$ and $ADFB'$. A 90° clockwise rotation about R takes $ABED'$ to $ADFB'$

and the segment RQ to RS. Hence QR and RS are equal in length and perpendicular. A similar $90°$ counterclockwise rotation about T shows that QT and TS are equal in length and perpendicular, from which it follows that $QRST$ is a square.

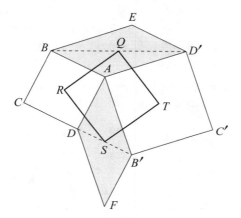

Figure S1.2.

1.7. Triangles ACF and BCE are congruent, and a $90°$ turn takes one to the other. Hence AF and BE are perpendicular at P. Since $\angle APE$ is a right angle, it lies on the circumcircle of square $ACED$. Hence $\angle DPE = 45°$ as it subtends a quarter circle. Similarly $\angle FPG = 45°$, so that $\angle DPG = 45° + 90° + 45° = 180°$, i.e., D, P, and G are collinear [Honsberger, 2001].

1.8. Yes, but only in a very special case. Let the original triangle have sides a, b, c and angles α, β, γ. Since each vertex angle of a regular polygon with n sides measures $(n-2)180°/n$ we need

$$360° - \frac{2(n-2)180°}{n} - \max\{\alpha, \beta, \gamma\} > 0,$$

or $\max\{\alpha, \beta, \gamma\} < 720°/n$. Hence $n \le 12$. If each flank triangle has the same area as the original triangle, then $bc \sin\alpha/2 = bc \sin(720°/n - \alpha)/2$, i.e., $\sin\alpha = \sin(720°/n - \alpha)$ (and similarly for β and γ). Hence $n = 4$, the classical configuration with squares, or $\alpha = \beta = \gamma = 360°/n$ and $n = 6$, i.e., regular hexagons around an equilateral triangle.

Chapter 2

2.1. We have equality in (2.1) if and only if $\sin\theta = 1$, i.e., $\theta = 90°$, in Figure 2.5. This implies that the shaded triangles are isosceles right triangles and thus $\sqrt{a} = \sqrt{b}$, so that $a = b$. We have equality in (2.2) if and only if $\theta = 90°$ in Figure 2.6 and $|ax + by| = |ax| + |by|$, i.e., ax and by have the same sign. This implies that the shaded triangles are similar, so that $|x|/|y| = |a|/|b|$, or $|ay| = |bx|$. If ax and by have the same sign, we have $ay = bx$.

2.2. The area of the white parallelogram in Figure S2.1a is $\sin(\pi/2 - \alpha + \beta)$ and $\sin(\pi/2 - \alpha + \beta) = \cos(\alpha - \beta)$ [Webber and Bode, 2002].

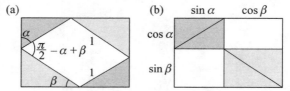

Figure S2.1.

2.3. Set $x = \sin t$ and $y = \cos t$ in Figure 2.5.

2.4. Using (2.2) we have

$$|ax + by + cz| \leq |ax + by| + |cz|$$
$$\leq \sqrt{a^2 + b^2}\sqrt{x^2 + y^2} + |cz|$$
$$\leq \sqrt{(a^2 + b^2) + c^2}\sqrt{(x^2 + y^2) + z^2}.$$

We can similarly prove by induction the n-variable Cauchy-Schwarz inequality

$$|a_1x_1 + a_2x_2 + \cdots + a_nx_n|$$
$$\leq \sqrt{a_1^2 + a_2^2 + \cdots + a_n^2}\sqrt{x_1^2 + x_2^2 + \cdots + x_n^2}.$$

2.5. Set x and y equal to $1/2$ in (2.2):

$$\frac{a+b}{2} = \left|a \cdot \frac{1}{2} + b \cdot \frac{1}{2}\right| \leq \sqrt{a^2 + b^2}\sqrt{\frac{1}{2}} = \sqrt{\frac{a^2 + b^2}{2}}.$$

2.6. No. Consider any isosceles triangle with $\angle A = \angle B$, $\angle C \neq 90°$.

Chapter 3

3.1. Use (3.1) with $a = |\sin \theta|$ and $b = |\cos \theta|$.

3.2. See Figure S3.1. The right triangles with legs a, $\sqrt{ab}$ and $\sqrt{ab}$, b are similar, and thus the shaded triangle is a right triangle. Its hypotenuse $a + b$ is at least as long as the base $2\sqrt{ab}$ of the trapezoid, thus $(a + b)/2 \geq \sqrt{ab}$.

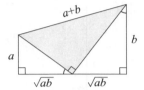

Figure S3.1.

3.3. See Figure S3.2.

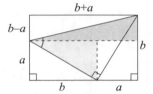

Figure S3.2.

3.4. Let $p = F_{2n}$ and $q = F_{2n-1}$ in (3.4). Then $p + q = F_{2n+1}$ and $p^2 + pq + 1 = F_{2n}^2 + F_{2n} F_{2n-1} + 1$. By Cassini's identity ($F_{k-1} F_{k+1} - F_k^2 = (-1)^k$ for $k \geq 2$), $F_{2n}^2 + 1 = F_{2n-1} F_{2n+1}$, and thus $p^2 + pq + 1 = F_{2n-1}(F_{2n} + F_{2n+1}) = q \cdot F_{2n+2}$. Hence $q/(p^2 + pq + 1) = 1/F_{2n+2}$.

3.5. See Figure S3.3.

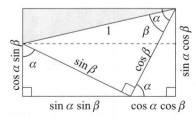

Figure S3.3.

3.6. See Figure S3.4.

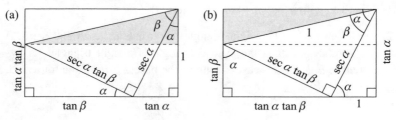

Figure S3.4.

3.7. See Figure S3.5 [Burk, 1996].

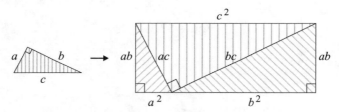

Figure S3.5.

3.8. Compute the lengths of the hypotenuse of each right triangle, as shown in Figure S3.6. Then in the shaded right triangle we have $\sin\theta = 2z/(1+z^2)$ and $\cos\theta = (1-z^2)/(1+z^2)$[Kung, 2001].

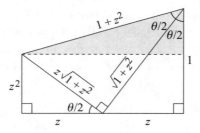

Figure S3.6.

3.9. If we place the trapezoid in the first quadrant, as shown in Figure S3.7, then the coordinates of P_k are

$$P_k = ((b+ak)/2, (a+bk)/2) = (b/2, a/2) + k(a/2, b/2),$$

so P_k lies on the ray from $(b/2, a/2)$ with slope b/a.

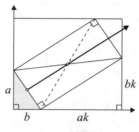

Figure S3.7.

3.10. In the Garfield trapezoid in Figure S3.8 we have $\sqrt{2(x + 1/x)} \geq \sqrt{x} + \sqrt{1/x}$, from which the first inequality in (3.5) follows. The second follows from the AM-GM inequality.

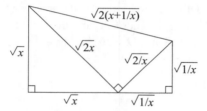

Figure S3.8.

Chapter 4

4.1. Draw AD and BD parallel to BC and AC, respectively, intersecting in D. Since C is a right angle, $ACBD$ is a rectangle. Hence its diagonals AB and CD are equal in length and bisect each other at O. Since O is equidistant from A, B, and C, it is the center of the circumcircle of triangle ABC, with AB as diameter. See Figure S4.1.

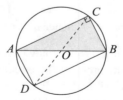

Figure S4.1.

4.2. Let $|PQ| = a$, $|QR| = b$, and $|QS| = h$, so that $h^2 = ab$ follows from the right triangle altitude theorem. If A denotes the area of the arbelos and C the area of the circle with diameter QS, then

$$A = \frac{\pi}{2}\left(\frac{a+b}{2}\right)^2 - \frac{\pi}{2}\left(\frac{a}{2}\right)^2 - \frac{\pi}{2}\left(\frac{b}{2}\right)^2 = \pi\left(\frac{ab}{4}\right) = \pi\left(\frac{h}{2}\right)^2 = C.$$

4.3. (a) Set $\theta = \arcsin x$, $\sin\theta = x$, and $\cos\theta = \sqrt{1-x^2}$ in Figure 4.14 and in the two half-angle tangent formulas. (b) Set $\theta = \arccos x$, $\cos\theta = x$, and $\sin\theta = \sqrt{1-x^2}$ in Figure 4.14 and in the two half-angle tangent formulas. (c) Set $\theta = \arctan x$, $\sin\theta = x/\sqrt{1+x^2}$, and $\cos\theta = 1/\sqrt{1+x^2}$ in Figure 4.14 and in the two half-angle tangent formulas.

4.4. With the solid line as the altitude, the area of the triangle is $(1/2)\sin 2\theta$. With the dashed line as the altitude, its area is $(1/2)(2\sin\theta)\cos\theta$, and hence $\sin 2\theta = 2\sin\theta\cos\theta$. Using the law of cosines, the length $2\sin\theta$ of the chord satisfies $(2\sin\theta)^2 = 1^2 + 1^2 - 2\cos 2\theta$, and hence $\cos 2\theta = 1 - 2\sin^2\theta$.

4.5. See Figure S4.2.

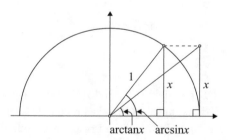

Figure S4.2.

4.6. The area of the shaded region is $\pi(R^2 - r^2)/2$. However, a is the geometric mean of $R+r$ and $R-r$, hence $a^2 = (R+r)(R-r) = R^2 - r^2$.

4.7. Let T denote the area of the right triangle, L_1 and L_2 the areas of the lunes, and S_1 and S_2 the areas of the circular segments (in white) in Figure S4.3. Then the version of the Pythagorean theorem employed in the proof of Proposition 4 from Archimedes' *Book of Lemmas* yields $T + S_1 + S_2 = (S_1 + L_1) + (S_2 + L_2)$, so $T = L_1 + L_2$.

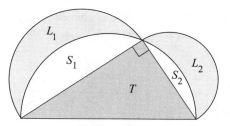

Figure S4.3.

4.8. We first compute the lengths of the segments in Figure 4.18 using triangle geometry: Let $|AB| = d$, then

$$|AF| = d\sqrt{6}/3, |BF| = d\sqrt{3}/3,$$
$$|AE| = d\sqrt{2}/2, \text{ and } |BN| = d/\phi\sqrt{3}.$$

(a) For a tetrahedron with edge s, consider the cross-section passing through one edge and the altitudes of the opposite faces, as shown in Figure S4.4.

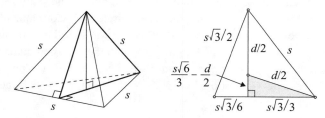

Figure S4.4.

Since the two altitudes have length $s\sqrt{3}/2$, s and d are related as shown in the gray right triangle, we have

$$\left(\frac{s\sqrt{6}}{3} - \frac{d}{2}\right)^2 + \left(\frac{s\sqrt{3}}{3}\right)^2 = \left(\frac{d}{2}\right)^2$$

and thus $s = d\sqrt{6}/3 = |AF|$.

(b) For a cube with edge s, the diameter d of the sphere is also the diagonal of an s-by-$s\sqrt{2}$ rectangle, so that $d^2 = s^2 + 2s^2$ and hence $s = d\sqrt{3}/3 = |BF|$.

(c) For an octahedron with edge s, d is the diagonal of a square with side s and hence $s = d\sqrt{2}/2 = |AE|$.

(d) For a dodecahedron with edge s, consider a cross-section passing through a pair of opposite edges, as shown in Figure S4.5.

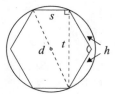

Figure S4.5.

The cross-section is a hexagon whose remaining four edges are altitudes of the pentagonal faces, of length $h = (s \tan 72°)/2$. The angle between two such edges is a dihedral angle $\arccos(-1/\sqrt{5})$ of the dodecahedron. If we let t be the length of the indicated segment in the figure, then the law of cosines along with identities such as $\sec 72° = 2\phi$ and $\phi + (1/\phi) = \sqrt{5}$ yields $t^2 = s^2(1+\phi)^2$. Hence we have $d^2 = s^2 + s^2(1+\phi)^2 = 3s^2(1+\phi) = 3s^2\phi^2$, or $s = d/\phi\sqrt{3} = |BN|$.

4.9. Assume without loss of generality that the radius of the small semicircles is 1 and the radius of the large semicircle is 2. Consequently the perimeter of the cardioid is 4π. It suffices to consider lines through the cusp that make an angle of θ in $[0, \pi/2]$ with the common diameter of the semicircles, as shown in Figure S4.6. The length of the portion of perimeter to the right of the line is $2\theta + 2(\pi - \theta) = 2\pi$, half the total perimeter.

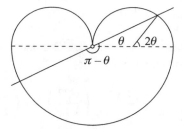

Figure S4.6.

Chapter 5

5.1. Since (a, b, c) and (a', b', c') are similar, $a' = ka$, $b' = kb$, and $c' = kc$ for some positive k. Hence

$$aa' + bb' = ka^2 + kb^2 = kc^2 = cc'.$$

5.2. Let s and t be the lengths of the segments indicated in Figure S5.1. Since each of the shaded triangles is similar to ABC, $b'/b = s/c$ and $a'/a = t/c$. Hence

$$\frac{a'}{a} + \frac{b'}{b} + \frac{c'}{c} = \frac{t}{c} + \frac{s}{c} + \frac{c'}{c} = \frac{c}{c} = 1$$

as claimed [Konhauser et al., 1996].

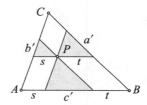

Figure S5.1.

5.3. Draw the diameter PS and the segment QS as shown in Figure S5.2. Since $\angle QPS = \angle PQR$, right triangle PQS is similar to right triangle PQR. Thus $|QR|/|PQ| = |PQ|/|PS|$, so that $|PQ|$ is the geometric mean of $|QR|$ and $|PS|$.

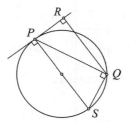

Figure S5.2.

5.4. See Figure S5.3. Draw KE parallel to CD. Triangles AKE and AFC are similar, thus $|FC|/|KE| = |AC|/|AE| = 2$, so that

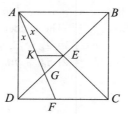

Figure S5.3.

$|FC| = 2|KE|$. Since $x = 22.5°$, $\angle DFG = 67.5° = \angle DGF$. Triangles DFG and EGK are similar, and since DFG is isosceles, so is EKG. Thus $|GE| = |KE|$, so that $|FC| = 2|GE|$.

5.5. Since the triangles have the angle at D in common, it is sufficient (and necessary) to have the angles marked x and y equal. Hence AD must be the bisector of the angle at A. See Figure S5.4.

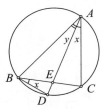

Figure S5.4.

5.6. In Figure S5.5ab we see that the I and P pentominoes are rep-4 reptiles. Since the L pentomino tiles a 2×5 rectangle (see Figure S5.5c) and the Y pentomino tiles a 5×10 rectangle (see Figure S5.5d), each tiles a 10×10 square and is thus a rep-100 reptile.

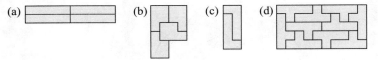

Figure S5.5.

5.7. No. Condition (5.3) yields $e^{bx} = ke^{ax/k}$ so that at $x = 0$ we have $k = 1$, i.e., $a = b$.

Chapter 6

6.1. From Figure S6.1 we have $2m_a \leq b+c, 2m_b \leq c+a, 2m_c \leq a+b$, so that $m_a + m_b + m_c \leq 2s$. The shaded triangle has sides $2m_a$, $2m_b$, $2m_c$ and medians $3a/2$, $3b/2$, $3c/2$, so by the above inequality we have $3(a + b + c)/2 \leq 2(m_a + m_b + m_c)$, or $3s/2 \leq m_a + m_b + m_c$.

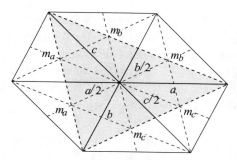

Figure S6.1.

6.2. See Figure S6.2a, where $a > b > c$. Since the sides are in arithmetic progression, $a + c = 2b$ so the semiperimeter is $s = 3b/2$. Hence the area K satisfies $K = rs = 3br/2$ and $K = bh/2$, thus $h = 3r$ so that the incenter I lies on the dashed line parallel to AC one-third of the way from AC to B (see Figure S6.2b). The centroid G lies on the median BM_b one-third of the way from AC to B, hence G also lies on the dashed line [Honsberger, 1978].

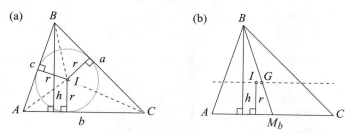

Figure S6.2.

6.3. Draw an altitude to one side, cut along two medians from the foot of the altitude in the resulting small triangles, and rotate the pieces as shown in Figure S6.3. Then rotate the entire new triangle 180° [Konhauser et al., 1996].

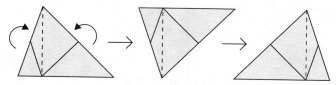

Figure S6.3.

6.4. Our solution is from [Erdös, 1940]. Since a cevian is shorter than either of the sides adjacent to the vertex from which it is drawn, it is shorter than the longest side of the triangle. Draw PR and PS parallel, respectively, to AC and BC as shown in Figure S6.4, so that $\triangle PRS$ is similar to $\triangle ABC$. Since PZ is a cevian in $\triangle PRS$ and RS is its longest side, $|PZ| < |RS|$.

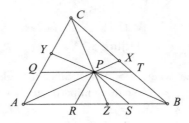

Figure S6.4.

Draw QT through P parallel to AB. $\triangle QPY$ is similar to $\triangle ABY$, and since AB is the longest side of $\triangle ABY$, QP is the longest side of $\triangle QPY$. Since $AQPR$ is a parallelogram, we have $|PY| < |QP| = |AR|$. Similarly, $|PX| < |PT| = |SB|$, and hence

$$|PX| + |PY| + |PZ| < |SB| + |AR| + |RS| = |AB|.$$

6.5. Triangles AHY and BHX in Figure 6.7 are similar, and hence $|AH|/|HY| = |BH|/|HX|$, so that $|AH| \cdot |HX| = |BH| \cdot |HY|$. In the same way $|BH| \cdot |HY| = |CH| \cdot |HZ|$.

6.6. Since the sides of $\triangle XYZ$ are parallel to the sides of $\triangle ABC$, the altitudes of $\triangle XYZ$ are perpendicular to the sides of $\triangle ABC$. See Figure S6.5, and let O be the orthocenter of $\triangle XYZ$. Thus $\triangle AOZ$ is congruent to $\triangle BOZ$, and $|AO| = |BO|$. Similarly $|BO| = |CO|$, whence O is equidistant from A, B, and C.

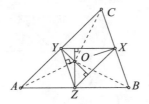

Figure S6.5.

6.7. See Figure S6.6, and let M be the midpoint of CD. Thus ME is parallel to BC. Since BC is perpendicular to AC, so is ME, hence it lies on the altitude to AC in $\triangle ACE$. In $\triangle ACE$, CD is an altitude, so M is the orthocenter of $\triangle ACE$. Thus AM lies on the third altitude, so AM is perpendicular to CE. Since A and M are midpoints of sides of $\triangle CDF$, AM is parallel to FD, and so FD is also perpendicular to CE [Honsberger, 2001].

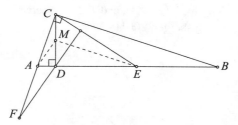

Figure S6.6.

6.8. (a) Draw PQ and CR perpendicular to AB, as shown in Figure S6.7.

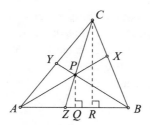

Figure S6.7.

Then right triangles PQZ and CRZ are similar, and so

$$\frac{|PZ|}{|CZ|} = \frac{|PQ|}{|CR|} = \frac{|PQ| \cdot |AB|/2}{|CR| \cdot |AB|/2} = \frac{[ABP]}{[ABC]}.$$

Similarly,

$$\frac{|PX|}{|AX|} = \frac{[BCP]}{[ABC]} \text{ and } \frac{|PY|}{|BY|} = \frac{[CAP]}{[ABC]}.$$

Hence

$$\frac{|PX|}{|AX|} + \frac{|PY|}{|BY|} + \frac{|PZ|}{|CZ|} = \frac{[ABP] + [BCP] + [CAP]}{[ABC]} = \frac{[ABC]}{[ABC]} = 1.$$

(b) Since $|PA|/|AX| = 1 - |PX|/|AX|$, etc., we have

$$\frac{|PA|}{|AX|} + \frac{|PB|}{|BY|} + \frac{|PC|}{|CZ|} = 3 - \left(\frac{|PX|}{|AX|} + \frac{|PY|}{|BY|} + \frac{|PZ|}{|CZ|} \right) = 3 - 1 = 2.$$

6.9. See Figure S7.8. Since X and Y are midpoints of BC and AC, XY is parallel to AB and half its length, or $|AZ| = |XY| = |BZ|$. Since PQ is parallel to AB and XY, it follows that PR, RS, and SQ are each half the length of AZ, XY, and BZ, respectively, hence $|PR| = |RS| = |SQ|$ [Honsberger, 2001].

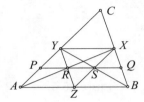

Figure S6.8.

6.10. We need only show that the semiperimeter s of a triangle of area $1/\pi$ exceeds 1. If r denotes the inradius of the triangle, we have $rs = 1/\pi$, or $s = 1/\pi r$. The area πr^2 of the incircle is less than the area of the triangle, so $\pi r^2 < 1/\pi$, so $1 < 1/(\pi r)^2$, or $1 < 1/\pi r$. Hence $s = 1/\pi r > 1$ [Honsberger, 2001].

Chapter 7

7.1. See Figure S7.1. Equality holds in (a) and (b) if and only if $ay - bx = 0$. i.e., $a/b = x/y$, and in (c) if and only if $a/b = x/y$ or $-y/x$.

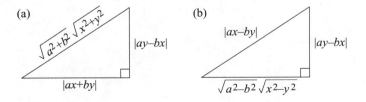

Figure S7.1.

7.2. See Figure S7.2.

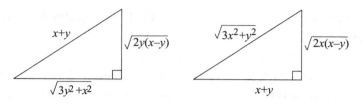

Figure S7.2.

7.3. (a) Two nonintersecting circles have two internal tangents that inter-sect at a point between the circles on the center line, the line joining the centers. Hence vertex A, the intersection of the lines containing side b and side c, lies on the line joining I_b and I_c; and similarly for vertices B and C.

(b) Two nonintersecting circles also have two external tangents that intersect at a point to one side of both circles on the center line. Furthermore, the configuration of two circles and four tangent lines is symmetric with respect to the center line. So if one internal and one external tangent are perpendicular, so are the other internal and external tangents. The excircles on sides a and c have the perpendicular lines of the legs as internal and external tangents, and hence the left-most dashed line is perpendicular to the line of the hypotenuse. Similar consideration of the a excircle and the incircle, the b excircle and the incircle, and the b and c excircles shows that the other three dashed lines are also perpendicular to the line of the hypotenuse and thus parallel to one another.

7.4. Yes. We have $(z + 2)^2 + [(2/z) + 2]^2 = [z + 2 + (2/z)]^2$ and if z is rational, so are all three sides.

7.5. (a) The length of the overlap is $a + b - c = 2r$.

(b) With sides a, b, c and altitude d, we have, applying part (a) of the problem to each of the three right triangles,

$$2r + 2r_1 + 2r_2 = (a + b - c) + (|AD| + h - b) + (|BD| + h - a)$$
$$= 2h + |AD| + |BD| - c = 2h,$$

and hence $r + r_1 + r_2 = h$ [Honsberger, 1978].

7.6. (a) Since $c = 2R = a + b - 2r$, $R + r = (a + b)/2 \geq \sqrt{ab} = \sqrt{2K}$.

(b) Since $rs = K = ch/2 \leq (2R \cdot R)/2 = R^2$, $R^2 - rs \geq 0$. But $s = 2R + r$ implies $rs = 2rR + r^2$, hence $R^2 - 2Rr - r^2 \geq 0$, or $(R/r)^2 - 2(R/r) - 1 \geq 0$. Because $R/r > 0$, R/r is at least as large as the positive root $1 + \sqrt{2}$ of $x^2 - 2x - 1 = 0$. Equality holds when the triangle is isosceles.

7.7. If we let $c = 1 - \sqrt{xy}$ and $b = \sqrt{(1-x)(1-y)}$, then $a = \sqrt{c^2 - b^2} = \sqrt{x + y - 2\sqrt{xy}}$. By the arithmetic-geometric mean inequality a is real so a, b, and c form a right triangle with $b \le c$; $a = 0$ if and only if $x = y$.

7.8. Introduce a coordinate system with the origin at C, AC on the positive x-axis, and BC on the positive y-axis, so that the coordinates of the points of interest are $M(b/2, 0)$, $N(0, r_a)$, and $I(r, r)$. Writing the equation for the line MN and using (7.2), it follows that I lies on MN.

7.9. Let a, b, c be the sides of the triangle, and s the side of the square, as shown in Figure S7.3.

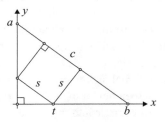

Figure S7.3.

Using similar triangles, $s = tc/b$, $s/(b-t) = a/c$, and hence $s = abc/(ab + c^2)$. If (a, b, c) is a primitive Pythagorean triple, then abc and $ab + c^2$ are relatively prime and s is not an integer. If (a, b, c) is replaced by (ka, kb, kc) where k is a positive integer, then $s = kabc/(ab + c^2)$ which will be an integer if and only if k is a multiple of $ab + c^2$. Thus the smallest such triangle similar to the primitive triangle is obtained by setting $k = ab + c^2$. Since the smallest (in terms of area and perimeter) primitive Pythagorean triple is $(3, 4, 5)$, and since it minimizes $ab + c^2$, the desired triple is obtained by multiplying by $3 \cdot 4 + 5^2 = 37$ to get $(111, 148, 185)$ [Yocum, 1990].

7.10. The bowtie lacing (d) is the shortest and the zigzag lacing (c) is the longest. We will show that the length order of the lacings is (d) < (a) < (b) < (c) by comparing edge lengths in right triangles whose vertices are the eyelets. Removing lace segments of equal length from the (d) and (a) lacings yields Figure S7.4. Comparing a leg and hypotenuse of a right triangle yields (d) < (a).

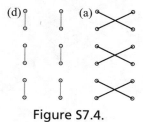

Figure S7.4.

Removing lace segments of equal length from the (a) and (b) lacings yields the patterns on the left in Figure S7.5, which are four copies of the simpler patterns in the center of the figure. On the right we rearrange the lace segments and use the fact that a straight line is the shortest path between two points to conclude (a) < (b).

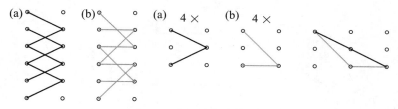

Figure S7.5.

Repeating the above procedure with lacings (b) and (c) yields Figure S7.6, and hence (b) < (c).

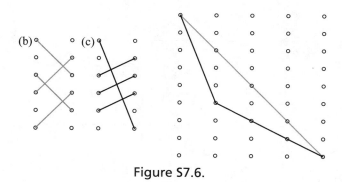

Figure S7.6.

7.11. See Figure S7.7, and compute $\sin\theta$ and $\cos\theta$ using the shaded triangle.

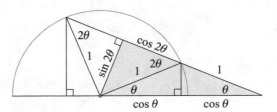

Figure S7.7.

7.12. See Figure S7.8.

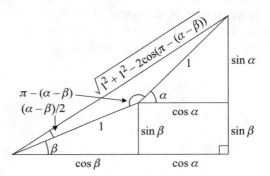

Figure S7.8.

7.13. The mean $xy\sqrt{(x+y)/(x^3+y^3)}$ is smaller than the harmonic mean, and the mean $\sqrt{(x^3+y^3)/(x+y)}$ is larger than the contraharmonic mean. Thus we can add two rows to Table 7.1:

	c	b	a
0)	$\dfrac{2xy}{x+y}$	$xy\sqrt{\dfrac{x+y}{x^3+y^3}}$	$\dfrac{\sqrt{3}xy\,|x-y|}{\sqrt{(x^3+y^3)(x+y)}}$
5)	$\sqrt{\dfrac{x^3+y^3}{x+y}}$	$\dfrac{x^2+y^2}{x+y}$	$\dfrac{|x-y|\sqrt{xy}}{x+y}$

Chapter 8

8.1. It suffices to show that $\angle PGR = 60°$ and $|PG| = 2|GR|$. Let T be the midpoint of BC. We first show that triangles PCG and RTG are similar. See Figure S8.1.

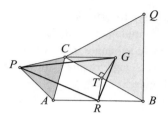

Figure S8.1.

Since RT is parallel to AC, $|PC| = |AC| = 2|RT|$. Since G is also the circumcenter of equilateral $\triangle BQC$, GT is perpendicular to BC, $\triangle CGT$ is a 30°-60°-90° triangle, so $\angle GCT = 30°$ and $|CG| = 2|GT|$. But $\angle RTB = \angle ACB$, so that $\angle RTG = \angle ACB + 90°$. However, $\angle PCG = 60° + \angle ACB + 30°$, hence triangles PCG and RTG are similar, and since $|CG| = 2|GT|$ we have $|PG| = 2|GR|$. Finally, $\angle PGR = \angle PGT + \angle TGR = \angle PGT + \angle PGC = 60°$, so $\triangle PGR$ is a 30°-60°-90° triangle, as claimed [Honsberger, 2001].

8.2. (a) Since X and Z are the midpoints of AC and CD, XZ is parallel to AD and $|XZ| = (1/2)|AD|$. Similarly YZ is parallel to BC and $|YZ| = (1/2)|BC|$. Since $|AD| = |BC|$, we have $|XZ| = |YZ|$. Because of parallels, $\angle YXZ + \angle XYZ = 120°$ and thus $\angle XZY = 60°$. Hence $\triangle XYZ$ is equilateral.

(b) Since $\angle A + \angle B = 120°$, then $\angle C + \angle D = 240°$. Thus

$$\angle PCB = 360° - \angle C - 60° = 300° - (240° - \angle D)$$
$$= 60° + \angle D = \angle ADP.$$

So triangles ADP and BCP are congruent and hence the angle between AP and BP is 60°. Thus $\triangle APB$ is equilateral [Honsberger, 1985].

8.3. The triangle with dashed sides in Figure 8.14 is the Napoleon triangle, since the outer vertex of a small gray triangle is the centroid of the equilateral triangle erected on the entire side of the given triangle, as illustrated in Figure S8.2.

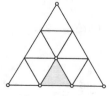

Figure S8.2.

8.4. Yes. There are two cases to consider in computing the area of a flank triangle. We use the formula $(xy \sin \theta)/2$ for the area of a triangle with sides x and y and included angle θ. When $\alpha > 60°$ equality of areas of ABC and the flank triangle requires $\sin \alpha = \sin(240° - \alpha)$, or $\alpha = 120°$, as shown in Figure S8.3a.

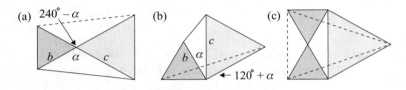

Figure S8.3.

When $\alpha < 60°$ equality of areas of ABC and the flank triangle requires $\sin \alpha = \sin(120° + \alpha)$, or $\alpha = 30°$, as shown in Figure S8.3b. The only triangle all of whose angles are 30° or 120° is a 30°–30°–120° triangle, as illustrated in Figure S8.3c.

8.5. Let a and b be the sides of $ABCD$, as shown in Figure S8.4. The gray triangles ABE and BCF are congruent, and hence $|BE| = |BF|$. Since the obtuse angles in the gray triangles are 150°, the two gray angles at vertex B sum to 30°, hence $\angle EBF = 60°$ and triangle BEF is equilateral, with side length $|BE| = |BF| = |EF| = \sqrt{a^2 + b^2 + ab\sqrt{3}}$ from the law of cosines. Let G and H denote the centroids of ADE and CDF and O the center of $ABCD$. Since $|GO| = (3a + b\sqrt{3})/6$ and $|HO| = (3b + a\sqrt{3})/6$, $|GH|^2 = |GO|^2 + |OH|^2 = (a^2 + b^2 + ab\sqrt{3})/3$, and thus $|GH| = |EF| \cdot \sqrt{3}/3$ as claimed.

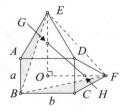

Figure S8.4.

8.6. Since triangles $BC'C$, $BA''C$, and $BB'C$ have the same base BC and equal angles C', A'', and B', they have the same circumcircle, as shown in Figure S8.5. Thus $\angle CA''B' = \angle B$ (in triangle ABC). Since $\angle BCA'' = \angle B + \angle C$ (in triangle ABC), $\angle C'CA'' = \angle C$ and hence line segments $A''B'$ and $C'C$ are parallel. Similarly $A''C'$ and $B'B$ are parallel, and hence $AC'A''B'$ is a parallelogram as claimed.

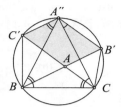

Figure S8.5.

8.7. See Figure S8.6 [Bradley, 1930].

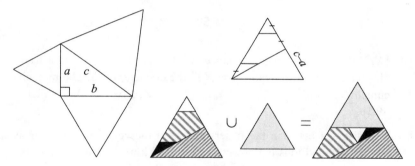

Figure S8.6.

Chapter 9

9.1. Computing the power of A with respect to the circle with $b = |AC|$ and $c = |AB|$ yields $(c - a)(c + a) = b^2$, or $c^2 = a^2 + b^2$.

9.2. Let AB, CD, and EF be the three chords, and set $a = |PA|$, $b = |PB|$, $c = |PC|$, $d = |PD|$, $e = |PE|$, and $f = |PF|$. Then $a + b = c + d = e + f$ and computing the power of P with respect to the circle yields $ab = cd = ef$. If we let s denote the common value of the sums and p the common value of the products, then each set $\{a, b\}$, $\{c, d\}$, $\{e, f\}$ are the roots of $x^2 - sx + p = 0$ and thus the sets are the same. Without loss of generality set $a = c = e$. Then the points A, C, and

E lie on a circle with center P and also on the given circle, so the two circles coincide and P is the center of the given circle [Andreescu and Gelca, 2000].

9.3. $\{A, X, Z_b, Y_c\}$ are concyclic since AXZ_b and AXY_c are right triangles with a common hypotenuse AX, the diameter of the common circumcircle; and similarly for $\{B, Y, X_c, Z_a\}$ and $\{C, Z, X_b, Y_a\}$. See Figure S9.1a.

(a)   (b)

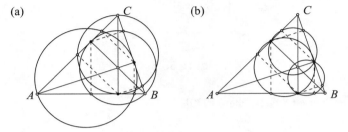

Figure S9.1.

9.4. $\{X, Y, X_c, Y_c\}$ are concyclic since XYX_c and XYY_c are right triangles with a common hypotenuse XY, the diameter of the common circumcircle. Similarly for $\{Y, Z, Y_a, Z_a\}$ and $\{Z, X, Z_b, X_b\}$. See Figure S9.1b.

9.5. The distance between the observatory and the center of the earth is 6382.2 km, and thus the power is 53,592.84 km². The distance to the horizon is the square root of the power, or about 231.5 km.

9.6. As usual, let a, b, c denote the sides opposite angles A, B, C, respectively. The power of A with respect to the circumcircle of $CBZY$ yields $c\,|AZ| = b\,|AY|$ and the power of B with respect to the circumcircle of $ACXZ$ yields $c\,|BZ| = b\,|BX|$. Adding yields $c^2 = a\,|BX| + b\,|AY|$. In both cases $|BX| = a - b\cos C$ and $|AY| = b - a\cos C$, and hence

$$c^2 = a(a - b\cos C) + b(b - a\cos C) = a^2 + b^2 - 2ab\cos C.$$

[Everitt, 1950].

9.7. See Figure S9.2, where two additional angles have measure y. Hence $\angle DOB = 2y$ and $\angle B = 90° - y$ since $\triangle DOB$ is isosceles. Since $ABDC$ is a cyclic quadrilateral, $x + (90° - y) = 180°$, or $x - y = 90°$.

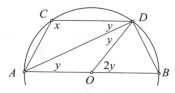

Figure S9.2.

Chapter 10

10.1. Rotate the inner square 45°, as shown in Figure S10.1.

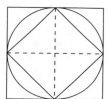

Figure S10.1.

10.2. Since Q is cyclic, Ptolemy's theorem $pq = ac + bd$ holds, and since Q is tangential, $a+c = b+d$. Hence two applications of the AM-GM inequality yield

$$8pq = 2(4ac + 4bd) \leq 2[(a+c)^2 + (b+d)^2] = (a+b+c+d)^2.$$

10.3. The centers of the two circles in Figure 10.20 lie on the bisector OC of $\angle AOB$, which is also the altitude from O in $\triangle AOB$ (see Figure S10.2). If we set $|OA| = 1$, then $|OA| = 1$ and $|OD| = \sqrt{2}/2$.

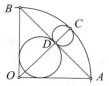

Figure S10.2.

Hence the diameter $|CD| = (2 - \sqrt{2})/2$. But from Challenge 7.5a the diameter of the incircle of $\triangle AOB$ is $|OD| = 2 - \sqrt{2} = 2|CD|$ [Honsberger, 2001].

10.4. Let segments $|OA|$, etc. be labeled as a, etc., and let K_1, etc. denote the areas of the shaded triangles in Figure S10.3. Since vertical angles at O are equal, as well as at A and C, and at B and D, we have

$$\frac{K_1}{K_2} = \frac{ae}{cg}, \frac{K_3}{K_4} = \frac{df}{bh}, \frac{K_1}{K_4} = \frac{ax}{by}, \text{ and } \frac{K_3}{K_2} = \frac{dx}{cy}.$$

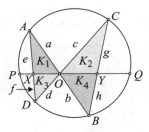

Figure S10.3.

Hence

$$\frac{K_1 K_3}{K_2 K_4} = \frac{adef}{bcgh} = \frac{adx^2}{bcy^2},$$

so that computing the powers of the points X and Y with respect to the circle yields

$$\frac{x^2}{y^2} = \frac{ef}{gh} = \frac{(p-x)(q+x)}{(p+y)(q-y)} = \frac{pq - x(q-p) - x^2}{pq + y(q-p) - y^2}.$$

Simplifying, $pq(y - x) = xy(q - p)$, from which the desired result follows [Bankoff, 1987].

10.5. See Figure S10.4, where we have drawn $ABCD$, and also the diameter DF and chord CF. Because each angle intercepts the same arc of the circle $\angle DAE = \angle DFC$, and $\triangle DCF$ is a right triangle.

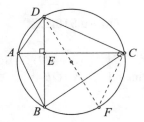

Figure S10.4.

Hence $\angle ADE = \angle FDC$, and thus $|AB| = |CF|$ as chords in equal arcs. Therefore

$$|AE|^2 + |BE|^2 + |CE|^2 + |DE|^2 = |AB|^2 + |CD|^2$$
$$= |CF|^2 + |CD|^2 = (2R)^2.$$

10.6. If $ABCD$ is a bicentric trapezoid, as shown in Figure S10.5a, then $\angle A + \angle C = \angle B + \angle D$ and $\angle A + \angle D = \angle B + \angle C(= 180°)$, and hence $\angle A = \angle B$ and $ABCD$ is isosceles. Then arc $AC = $ arc BD and hence arc $AD = $ arc BC. Thus $|AD| = |BC| = u$ (say), and let x and y denote the bases as shown in Figure S10.5b.

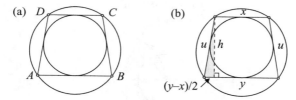

Figure S10.5.

Then $x + y = 2u$ so that $u = (x + y)/2$. Let h denote the altitude of $ABCD$. Applying the Pythagorean theorem to the gray triangle yields $h = \sqrt{xy}$.

Chapter 11

11.1. Draw line segments PQ, BP, and CQ, as shown in Figure S11.1. Since $\triangle APB$ and $\triangle AQC$ are isosceles and BP and CQ are parallel, we have

$$\angle BAC = 180° - (\angle PAB + \angle QAC)$$
$$= 180° - (1/2)(180° - \angle APB + 180° - \angle AQC)$$
$$= (1/2)(\angle APB + \angle AQC) = (1/2)(180°) = 90°.$$

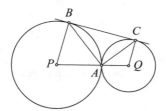

Figure S11.1.

11.2. Using angle measurement methods from Section 9.2, we have $\angle QPR = x/2 = (z - y)/2$, and hence $x + y = z$. It is not necessary for the circles to be tangent.

11.3. Draw AO intersecting the given circle at P as shown in Figure S11.2. Since AO bisects $\angle BAC$, we need only show that BP bisects $\angle ABC$ to conclude that P is the incenter of $\triangle ABC$. Because $\triangle OPB$ is isosceles, we have

$$\angle CBO + \angle PBC = \angle PBO = \angle OPB = \angle PAB + \angle APB.$$

But $\angle BPO = \angle PAB$ since corresponding sides are perpendicular, hence $\angle PBC = \angle APB$ and BP bisects $\angle ABC$.

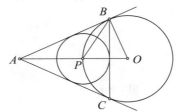

Figure S11.2.

11.4. Orient the lamina as shown in Figure S11.3. Computing its moment about the vertical axis of the crescent as the difference of the moments of the two disks yields

$$2x \cdot (\pi \cdot 1^2 - \pi x^2) = 1 \cdot (\pi \cdot 1^2) - x \cdot (\pi x^2).$$

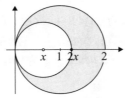

Figure S11.3.

Simplifying yields $x^3 - 2x + 1 = 0$, which has three real roots: 1, $-\phi$, and $1/\phi$, with only the last one being meaningful [Glaister, 1996].

11.5. Two roots ($\sqrt{1}$ and $\sqrt{4}$) are the radius and diameter of each of the circles. The others can be seen in Figure S11.4.

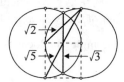

Figure S11.4.

11.6. Since the circular window lies between two arcs of radii (say) r and $2r$, its radius is $r/2$ and its center lies at the intersection of two arcs with radii $3r/2$. See Figure S11.5.

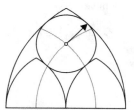

Figure S11.5.

11.7. Let $RQPST$ be a chord of the outer circle, as shown in Figure S11.6. The power of P with respect to each circle is constant, say $|PQ| \cdot |PS| = c$ and $|PR| \cdot |PT| = k$. Since $|ST| = |QR|$ we have

$$k = (|PQ| + |QR|)(|PS| + |ST|) = (|PQ| + |QR|)(|PS| + |QR|)$$
$$= |QR| \cdot (|QR| + |PQ| + |PS|) + c = |QR| \cdot (|QR| + |QS|) + c.$$

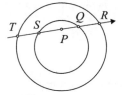

Figure S11.6.

Thus $|QR| \cdot (|QR| + |QS|)$ is constant, so $|QR|$ will be a maximum whenever $|QS|$ is a minimum. The arithmetic mean-geometric mean inequality applied to $|PQ|$ and $|PS|$ with $|PQ| \cdot |PS| = c$ tells us that $|QS|$ is a minimum when P is the midpoint of QS, and P bisects QS if and only if PR is perpendicular to the segment joining P to the common center of the two circles [Konhauser et al., 1996].

11.8. Since $\angle OMP$ is a right angle, the locus is the arc of the circle with diameter OP inside C.

11.9. Ratios of corresponding sides in similar triangles yields

$$\frac{|VP|}{r} = \frac{|VP| + r}{R} = \frac{|VP| + r + r'}{r'},$$

from which it follows that $R = 2rr'/(r + r')$.

11.10. From Figure S11.7, we have

$$|AB|^2 = (r_1 + r_2)^2 - (r_1 - r_2)^2 = 4r_1r_2$$

and the result follows.

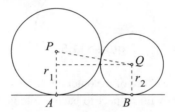

Figure S11.7.

11.11. Using the solution to Challenge 11.10, we have $|AC| = 2\sqrt{r_1r_3}$, $|CB| = 2\sqrt{r_2r_3}$, and $|AC| + |CB| = |AB| = 2\sqrt{r_1r_2}$. Hence $\sqrt{r_2r_3} + \sqrt{r_1r_3} = \sqrt{r_1r_2}$, and division by $\sqrt{r_1r_2r_3}$ yields the desired result.

Chapter 12

12.1. Using the Hint and the notation in Figure 12.5b, we have $A + B + C + D = T_1 + T_2 = T$.

12.2. The right angles at Q, R, and S enable us to draw circles through P, R, B, Q and through P, R, S, C as shown in Figure S12.1. Hence $\angle PRS + \angle PCS = 180°$. But $\angle PCS = 180° - \angle PBA =$

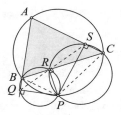

Figure S12.1.

$\angle PBQ = \angle PRQ$, and thus $\angle PRS + \angle PRQ = 180°$. Hence Q, R, and S lie on the same line.

12.3. View the intersecting circles as a graph with V vertices (points of intersection of the circles), E edges (arcs on the circles) and F faces (regions in the plane), and use Euler's formula: $V - E + F = 2$. Each circle intersects the others, so it will have $2(n-1)$ vertices and $2(n-1)$ edges. Hence $E = 2n(n-1)$. Since each vertex belongs to two circles, $V = n(n-1)$, and thus $F = 2 + 2n(n-1) - n(n-1) = n^2 - n + 2$.

12.4. Two. One is AB. The other is determined by PAQ where the circle C_3 is symmetric to C_2 with respect to the tangent line to C_2 at A. See Figure S12.2.

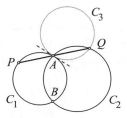

Figure S12.2.

12.5. Let K denote the area of the shaded region in Figure 12.15. Then $K = 2L + 4M - 1$ where L is the area of the lens-shaped region inside two quarter circles whose centers are at the endpoints of one of the diagonals of the square, and M is the area of one of the four regions adjacent to a side of the square. Hence $L = 2(\pi/4) - 1 = \pi/2 - 1$ and since M is the difference between the area of the square and the sum of the areas of two circular sectors (with angle $\pi/6$, i.e., area $\pi/12$) and an equilateral triangles with side 1, $M = 1 -$

$2(\pi/12) - \sqrt{3}/4 = 1 - \pi/6 - \sqrt{3}/4$. Hence $K = 1 + \pi/3 - \sqrt{3}$ $\approx$ 0.315.

12.6. The solution of the bun puzzle comes from the fact that, with the relative dimensions of the circles as given, the three diameters will form a right-angled triangle, as shown in Figure S12.3a. It follows that the two smaller buns are exactly equal in area to the large bun. Therefore, if we give David and Edgar the two halves of the largest bun, they will have their fair shares—one quarter of the confectionery each (Figure S12.3b). Then if we place the small bun on the top of the remaining one and trace its circumference in the manner shown in Figure S12.3c, Fred's piece (in gray) will exactly equal Harry's small bun with the addition of the white piece—half the rim of the other. Thus each boy gets an equal share, and only five pieces are necessary.

(a)   (b)   (c)

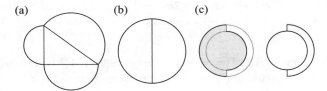

Figure S12.3.

12.7. Join the centers of the circles to form an equilateral triangle. Then the area of the shaded region in Figure 12.17 is the area $\sqrt{3}r^2$ of the triangle minus the sum $\pi r^2/2$ of the areas of three sectors, i.e., $(\sqrt{3} - \pi/2)r^2$.

12.8. Yes. See Figure S12.4.

Figure S12.4.

12.9. Triangle PAB will be isosceles if $|PA| = |PB|$, $|PA| = |AB|$, or $|PB| = |AB|$. Since $\triangle ABC$ is equilateral, Figure S12.5 shows that there are ten positions for the point P [Honsberger, 2004].

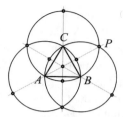

Figure S12.5.

12.10. Since the three angles at F in Figure 12.18 are each $120°$, and the angles at A', B', and C' are each $60°$, the circumcircles of the three shaded triangles all pass though F. Hence the line segment joining the centers of the circumcircles of $\triangle ABC'$ and $\triangle AB'C$ is perpendicular to the common chord AF, and similarly for the chords BF and CF.

Chapter 13

13.1. The result follows from the carpets theorem if the sum of the areas of quadrilaterals $AMCP$ and $BNDQ$ equals the area of $ABCD$. See Figure S13.1. The area of quadrilateral $AMCP$ equals the sum of the areas of triangles ACM and ACP. The area of triangle ACM is one-half the area of triangle ABC and the area of triangle ACP is one-half the area of triangle ACD, so the area of quadrilateral $AMCP$ is one-half of the area of quadrilateral $ABCD$. A similar result holds for quadrilateral $BNDQ$, and hence the sum of the areas of quadrilaterals $AMCP$ and $BNDQ$ equals the area of $ABCD$ [Andreescu and Enescu, 2004].

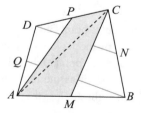

Figure S13.1.

13.2. To reach the conclusion we need only show that the sum of the areas of $PMQN$ and $PRQS$ equals the area of $ABCD$. Label the vertices of the three rectangles as shown in Figure S13.2, and draw the diagonals of $PMQN$ and $PRQS$ (the dashed lines), and let O be their common point of intersection. Since O is the midpoint of PQ, it lies on the vertical axis of symmetry of $ABCD$, and since O lies of MN it lies on the horizontal axis of symmetry of $ABCD$, and hence O is the center of $ABCD$. It now follows that $|DS| = |CN| = |AM|$, and thus SM is parallel to AD.

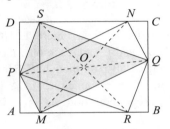

Figure S13.2.

The shaded quadrilateral $PMQS$ is the union of two triangles PMS and QSM. The area of PMS is one-half the area of $AMSD$ and the area of QSM is one-half the area of $MBCS$, so the area of $PMQS$ is one-half the area of $ABCD$. Similarly, $PMQS$ is the union of the two triangles PQM and PQS, and thus it is equal in area to one-half of $PMQN$ plus one-half of $PRQS$, and thus the area of $ABCD$ equals the sum of the areas of $PMQN$ and $PRQS$, as claimed [Konhauser et al., 1996].

13.3. In (a) we have $a/2 + b/2 \geq \sqrt{ab}$, the arithmetic mean-geometric mean inequality [Kobayashi, 2002]. In (b) we have $a^2/2 + b^2/2 \geq [(a + b)/2]^2$, which on taking square roots yields the arithmetic mean-root mean square inequality.

13.4. Three sets of four cocircular points are shown as black dots in Figure S13.3: the four corners of the isolated rectangle, the four points on the small white disk, and the points labeled A, B, C, and D. To see this draw BD and consider it as the diameter of a circle. Since the angles at A and C are right angles, they lie on the circle of which BD is the diameter.

There is a fifth set of four cocircular points: A, the unmarked intersection immediately below A on the right, B, and the unmarked corner just above B. The points all lie on the circle whose diameter is the line segment joining B to the point below A.

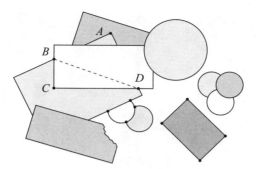

Figure S13.3.

13.5. *A* covers more than half the area of *B*—consider the area of the triangle whose vertices are the common corner of *A* and *B*, the right-most corner of *B*, and the left-most corner of *A*.

Chapter 14

14.1. Assume the radius of the monad is 1. In each case we need only show yin is bisected. (a) Yin is bisected since the area of a small circle is $\pi/4$. (b) The area inside the circular cut is $\pi/2$, and by symmetry the area of the part of yin inside the cut is $\pi/4$. (c) With the radii given in the Hint, the part of yin adjacent to the center of the monad has area

$$\frac{\pi}{8\phi^2} + \frac{\pi\phi^2}{8} - \frac{\pi}{8} = \frac{(2-\phi)\pi}{8} + \frac{(1+\phi)\pi}{8} - \frac{\pi}{8} = \frac{\pi}{4}$$

[Trigg, 1960]. (d) Use the procedure shown in Figure 14.3, but with four regions rather than seven.

14.2. See Figure S14.1 [Duval, 2007].

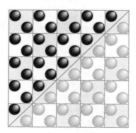

Figure S14.1.

14.3. See Figure S14.2 [Larson, 1985].

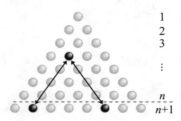

Figure S14.2.

14.4. Let $N = 2^n(2k + 1)$ where $n \geq 0$ and $k \geq 1$, and express $2N$ as the product of $m = \min\{2^{n+1}, 2K + 1\}$ and $M = \max\{2^{n+1}, 2K + 1\}$. Since $(M - m + 1)/2$ and $(M + m - 1)/2$ are positive integers whose sum is M, we can represent $2N$ by the array of balls shown in Figure S14.3:

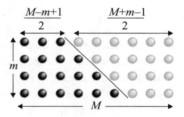

Figure S14.3.

Thus $N = \frac{M-m+1}{2} + \frac{M-m+3}{2} + \cdots + \frac{M+m-1}{2}$ [Frenzen, 1997].

14.5. Arrange three copies of $1^2 + 2^2 + 3^2 + \cdots + n^2$ cubes as shown in Figure S14.4a into the object shown in Figure 14.4b. Two copies it form a rectangular box with dimensions $n \times (n + 1) \times (2n + 1)$ as shown in Figure S14.4c, and the formula follows from division by 6.

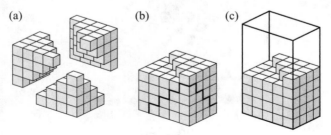

Figure S14.4.

14.6. Use symmetry to obtain (a) $\pi/2$, (b) $(\pi \ln 2)/8$, (c) 1, (d) 0, (e) 1/2, (f) π.

14.7. Draw a line from A internally tangent to the boundary of yin at T, and draw a line from the end of the diameter opposite A through T to intersect the boundary of yin at S. Then B can be located anywhere on the portion of the boundary between T and S passing through A. See Figure S14.5a.

(a)   (b)   (c)   (d)

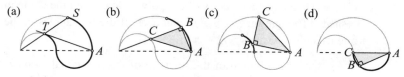

Figure S14.5.

In Figures 14.5b, c, and d we see three cases for the right angle at B and the location of vertex C.

Chapter 15

15.1. Let n be trapezoidal, specifically the sum of m consecutive positive integers beginning with $k + 1$. Then (with $T_0 = 0$)

$$n = (k + 1) + (k + 2) + \cdots + (k + m) = T_{k+m} - T_k$$
$$= \frac{(k + m)(k + m + 1)}{2} - \frac{(k)(k + 1)}{2} = \frac{m(2k + m + 1)}{2}.$$

One of m or $2k + m + 1$ is odd and the other is even, and hence if n is trapezoidal, it is not a power of 2 [Gamer et al., 1985].

15.2. Both recursions follow immediately from (15.1).

15.3. See Figure S15.1 [Nelsen, 2006].

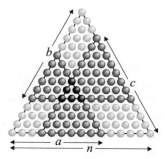

Figure S15.1.

15.4. In Section 15.5 we showed that the sum of the squares of the diagonals from a vertex and the two adjacent sides is $2nR^2$. Repeating
for each vertex counts the squares of all the diagonals and all the
sides twice. Hence the sum is $(1/2)n(2nR^2) = n^2R^2$ [Ouellette and
Bennett, 1979].

15.5. The two acute angles at the base each measure $\pi/2 - \pi/n$, and each
of the other $n-2$ obtuse angles measures $\pi - \pi/n$. See Figure 15.17b,
and observe that at each vertex of the polygonal cycloid the angle in
the shaded triangle measures π/n and each of the two angles in the
white triangle measures $\pi/2 - \pi/n$.

15.6. Let L_k denote the length of the segment of the polygonal cardioid
in the isosceles triangle with equal sides d_k in Figure 15.18b. Since
$d_k = 2R\sin(k\pi/n)$, we have

$$L_k = 4R\sin\frac{k\pi}{n}\sin\frac{2\pi}{n} = 2R\left[\cos\frac{(k-2)\pi}{n} - \cos\frac{(k+2)\pi}{n}\right]$$

and hence the length of the polygonal cardioid is

$$\sum_{k=1}^{n-1} L_k = 2R\sum_{k=1}^{n-1}\left[\cos\frac{(k-2)\pi}{n} - \cos\frac{(k+2)\pi}{n}\right]$$
$$= 4R\left(1 + 2\cos\frac{\pi}{n} + \cos\frac{2\pi}{n}\right) = 8R\cos^2\frac{\pi}{n} + 8r$$

$(r = R\cos(\pi/n))$.

15.7. If the angle of inclination of the chord is θ_0, then the length of the
chord is

$$2a(1 + \cos\theta_0) + 2a(1 + \cos(\theta_0 + \pi)) = 4a.$$

Chapter 16

16.1. See Figure S16.1, and apply the law of sines to the gray triangle to obtain $(a_2 + b_3)/(c_3 + d_1) = \sin D/\sin A$. Similarly,
we have $(b_2 + c_3)/(d_3 + e_1) = \sin E/\sin B$, $(c_2 + d_3)/(e_3 +
a_1) = \sin A/\sin C$, $(d_2 + e_3)/(a_3 + b_1) = \sin B/\sin D$, and
$(e_2 + a_3)/(b_3 + c_1) = \sin C/\sin E$. Multiplying the equations yields
(16.2) [Lee, 1998].

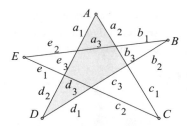

Figure S16.1.

16.2. Since area(CDE) = area(DEA), the line segment AC is parallel to DE. Similarly, each diagonal of the pentagon is parallel to a side. Thus $ABCF$ (see Figure S16.2a) is a parallelogram, and area(ACF) = area(ABC) = 1.

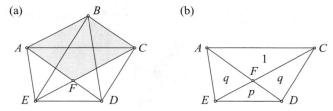

Figure S16.2.

Let p = area(DEF) and q = area(CDF) = area(AEF), with $p + q = 1$. Since triangles ACF and AEF have the same base AF and triangles CDF and DEF have the same base DF, $1/q = q/p$, or $p = q^2$ (see Figure S16.2b). Thus $q^2 + q = 1$ so that $q = \phi - 1$, $p = 2 - \phi$, and the area of the pentagon is $1 + 1 + p + 2q = \phi + 2$ [Konhauser et al., 1996].

16.3. (a) Erase some of the lines in Figure 16.23b and draw some others to create the grid in Figure S16.3a. Hence the triangle is a right triangle whose legs are in the ratio 3:4, and the result follows.

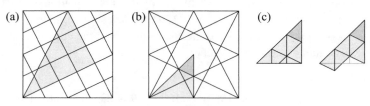

Figure S16.3.

(b) We need only show that the dark gray portion of the shaded right triangle in Figure S16.3b has $1/6$ the area of the entire shaded right triangle, and that is done in Figure S16.3c.

16.4. Yes, see Figure S16.4.

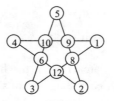

Figure S16.4.

16.5. We need to show that $a + b + c = x + y + z$ (See Figure S16.5).

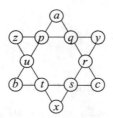

Figure S16.5.

Let N denote the magic constant. Computing the six line sums yields

$$a + p + u + b = N = x + r + s + y$$
$$b + t + s + c = N = y + p + q + z$$
$$c + r + q + a = N = z + u + t + x$$

from which the result follows by adding the three equations and canceling common terms (p, q, r, s, t, u) on each side. Similarly, in any magic octagram the sum of the four numbers at the corners of each large square must be the same.

16.6. See Figure S16.6.

The figure illustrates $r(9, 3) = 10$, and 9 is the smallest value of n for which $r(n, 3) > n$.

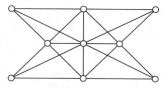

Figure S16.6.

16.7. Yes. See Figure S16.7, where ⭕ represents a standing tree and •
a harvested tree.

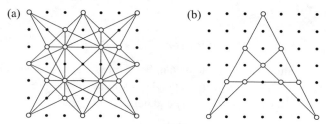

Figure S16.7.

16.8. In Figure S16.8 AB is a side of the decagon and FD a side of the
$\{10/3\}$ star. Draw diameters BE and GC through the center O. Since
the decagon is regular, the chords AB, GC, and FD are parallel, as
are the chords AF, BE, and CD. Hence the two shaded quadrilaterals
are parallelograms, so that

$$|FD| = |HC| = |HO| + |OC| = |AB| + |OC|$$

as claimed [Honsberger, 2001].

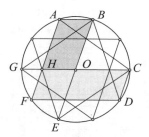

Figure S16.8.

16.9. See Figure S16.9.

(a)   (b)

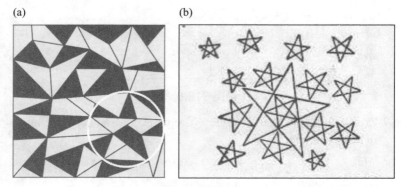

Figure S16.9.

16.10. Let the dimensions of the rectangle be x-by-1 as indicated in Figure S16.10. Since the entire rectangle is similar to the portion shaded gray, we have $x/1 = 1/(x/2)$ and hence $x = \sqrt{2}$.

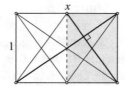

Figure S16.10.

Chapter 17

17.1. $\frac{1}{8} + \frac{1}{16} + \frac{1}{32} + \cdots = \frac{1}{4}$ and $\frac{2}{9} + \frac{2}{27} + \frac{2}{81} + \cdots = \frac{1}{3}$.

17.2. See Figure S17.1 for the first iteration, placing 25 non-attacking queens on a 25-by-25 chessboard [Clark and Shisha, 1988].

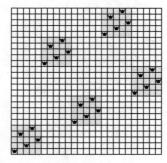

Figure S17.1.

17.3. Let the original triangle have edge length 1, and let A_n and L_n denote the area and length of the boundary after n iterations. Then $A_n = (3/4)A_{n-1}$ and $L_n = (3/2)L_{n-1}$ with $A_0 = \sqrt{3}/4$ and $L_0 = 3$, so that $A_n = (3/4)^n \cdot (\sqrt{3}/4)$, $L_n = 3(3/2)^n$, and the result follows.

17.4. Let A_n and P_n denote the area and the perimeter of the holes in the Sierpiński carpet after n iterations. It is easy to show that $A_n = (8/9)^n$ and $P_n = (4/5)[(8/3)^n - 1]$, so the result follows.

17.5. See Figure S17.2, where parts a, b, c, d illustrate $n = 2, 3, 4, 5$, respectively. In Figure S17.2a, the right triangle is isosceles, in S17.2b the acute angles are 30° and 60°, in S17.2c the right triangle is arbitrary, and in S17.2d the legs are 1 and 2.

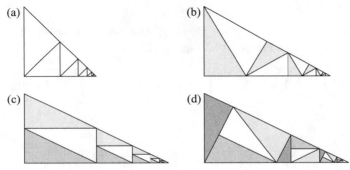

Figure S17.2.

17.6. Since $\phi - 1 = 1/\phi$, $1 - 1/\phi = 1/\phi^2$, and so on, the areas of the white squares in Figure 17.22c are 1, $1/\phi^2$, $1/\phi^4$, $1/\phi^6$, etc., which sum to the area ϕ of the original rectangle.

Chapter 18

18.1. $\frac{2}{9} + \frac{2}{81} + \frac{2}{729} + \cdots = \frac{1}{4}$.

18.2. (a) Let $a = F_n$ and $b = F_{n+1}$ so that $a + b = F_{n+2}$ and $b - a = F_{n-1}$.

(b) The identity in (a) is equivalent to $F_{n+1}^2 - F_n F_{n+2} = F_{n-1}F_{n+1} - F_n^2$. Thus the terms of the sequence $\{F_{n-1}F_{n+1} - F_n^2\}_{n=2}^{\infty}$ have the same magnitude and alternate in sign. So we need only evaluate the base case $n = 2$: $F_1 F_3 - F_2^2 = 1$ so $F_{n-1}F_{n+1} - F_n^2$ is $+1$ when n is even and -1 when n is odd, i.e., $F_{n-1}F_{n+1} - F_n^2 = (-1)^n$.

18.3. Figure 18.12 suffices, with α replaced by its complement in the light gray tatamis [Webber and Bode, 2002].

18.4. See Figure S18.1.

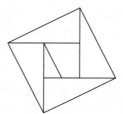

Figure S18.1.

18.5. In Figure S18.2a we see that $1 \geq 4pq$, so that $(1/p) + (1/q) = 1/pq \geq 4$. In Figure S18.2b we see that

$$2\left(p + \frac{1}{p}\right)^2 + 2\left(q + \frac{1}{q}\right)^2 \geq \left(p + \frac{1}{p} + q + \frac{1}{q}\right)^2 \geq (1+4)^2 = 25.$$

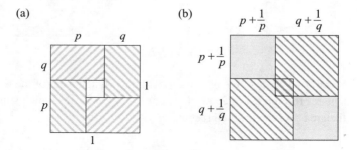

Figure S18.2.

18.6. Using the hint, we have

$$T_{8T_n} = \frac{8T_n(8T_n + 1)}{2} = 4T_n(2n + 1)^2,$$

and so if T_n is square, so is T_{8T_n}. Since $T_1 = 1$ is square, this relation generates an infinite sequence of square triangular numbers.

18.7. See Figure S18.3.

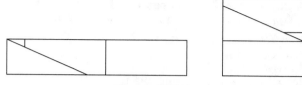

Figure S18.3.

18.8. See Figure S18.4, where in each case the side of the large square is $F_{n+1} = F_n + F_{n-1}$, and in Figures S18.4bc the side of the small central square is F_{n-2}.

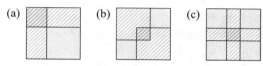

Figure S18.4.

Chapter 19

19.1. If $A = (a, 1/a)$ and $B = (b, 1/b)$, then $M = ((a+b)/2, ((1/a) + (1/b))/2)$ and if $P = (p, 1/p)$, then

$$\frac{1/p}{p} = \frac{((1/a) + (1/b))/2}{(a+b)/2} = \frac{1}{ab}.$$

Hence $p = \sqrt{ab}$ and $1/p = \sqrt{1/ab}$ as required [Burn, 2000].

19.2. The proof follows immediately from the observation that the two shaded triangles in Figure S19.1 have the same area.

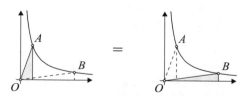

Figure S19.1.

19.3. (a) For $x > 0$, (19.2) implies $1/(1+x) < [\ln(1+x)]/x < 1$ and for x in $(-1,0)$, (19.2) implies $1 < [\ln(1+x)]/x < 1/(1+x)$. (b) Use (19.3) with $\{a, b\} = \{1, 1+x\}$.

19.4. Replacing x by $x^{1/n} - 1$ in part (a) of Challenge 19.3 yields $1 - x^{-1/n} < (1/n) \ln x < x^{1/n} - 1$, or $n(1 - x^{-1/n}) < \ln x < n(x^{1/n} - 1)$, which is equivalent to $\ln x < n(x^{1/n} - 1) < x^{1/n} \ln x$. Take the limit as $n \to \infty$.

19.5. Set $\{a, b\} = \{1 - x, 1 + x\}$ in (19.3).

19.6. Replace a by $\sqrt{a}$ and b by $\sqrt{b}$ in (19.3) to obtain

$$\sqrt[4]{ab} \le \frac{2(\sqrt{b} - \sqrt{a})}{\ln b - \ln a} \le \frac{\sqrt{a} + \sqrt{b}}{2}$$

and multiply by $(\sqrt{a} + \sqrt{b})/2$ to obtain the two inner inequalities in (19.4). The outer inequalities are equivalent to the AM-GM inequality [Carlson, 1972].

Chapter 20

20.1. Yes. See Figure S20.1 for examples from the Real Alcázar in Seville.

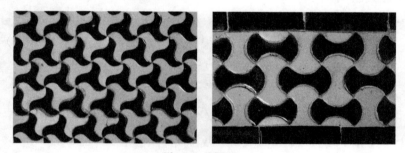

Figure S20.1.

20.2. See Figure S20.2.

Figure S20.2.

20.3. This is Bhāskara's proof. See Section 18.1.

20.4. Using the Hint, we see that the area of the smaller square is 4 and the
area of the larger square is 10, hence the smaller square has 2/5 the
area of the larger. See Figure S20.3.

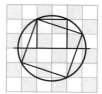

Figure S20.3.

20.5. Using the Hint (see Figure S20.4), we need only find the area of
the central triangle, since all four triangles in the Vecten configuration have the same area. Enclosing the central triangle in the 20-acre
dashed rectangle yields its area: $20 - (5/2 + 9/2 + 4) = 9$ acres.
Hence the area of the estate is $26 + 20 + 18 + 4 \times 9 = 100$ acres.

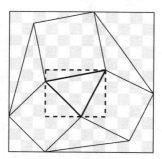

Figure S20.4.

20.6. In Figure S20.5 we see that the sum $|a||x| + |b||y|$ of the area of the
two rectangles is the same as the area of a parallelogram with edge
lengths $\sqrt{a^2 + b^2}$ and $\sqrt{x^2 + y^2}$.

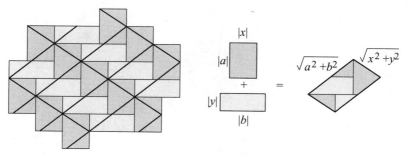

Figure S20.5.

The area of a parallelogram with given edges lengths is less that the area of a rectangle with the same edge lengths, and thus

$$|ax + by| \leq |a||x| + |b||y| \leq \sqrt{a^2 + b^2}\sqrt{x^2 + y^2}.$$

References

A. Aaboe, *Episodes From the Early History of Mathematics*. Random House, New York, 1964.

E. A. Abbott, *Flatland: A Romance of Many Dimensions*. Seeley & Co., London, 1884 (reprinted by Dover Publications, New York, 1992).

S. A. Ajose, Proof without words: Geometric series. *Mathematics Magazine*, 67 (1994), p. 230.

C. Alsina and R. B. Nelsen, *Math Made Visual: Creating Images for Understanding Mathematics*, Mathematical Association of America, Washington, 2006.

———, Geometric proofs of the Weitzenböck and Hadwiger-Finsler inequalities. *Mathematics Magazine*, 81 (2008), pp. 216–219.

———, *When Less is More: Visualizing Basic Inequalities*. Mathematical Association of America, Washington, 2009.

———, *Charming Proofs: A Journey Into Elegant Mathematics*, Mathematical Association of America, Washington, 2010.

T. Andreescu and B. Enescu, *Mathematical Olympiad Treasures*. Birkhäuser, Boston, 2004.

T. Andreescu and R. Gelca, *Mathematical Olympiad Challenges*. Birkhäuser, Boston, 2000.

T. M. Apostol and M. A. Mnatsakanian, Cycloidal areas without calculus. *Math Horizons*, September 1999, pp. 12–16.

J.-L. Ayme, La Figure de Vecten. http://pagesperso-orange.fr/jl.ayme/Docs/La figure de Vecten.pdf, accessed 6 June 2010.

H. H. R. Ball, *A Short Account of the History of Mathematics*. Macmillan and Co., London, 1919 (reprinted by Dover Publications, New York, 1960).

T. Banchoff and P. Giblin, On the geometry of piecewise circular curves, *American Mathematical Monthly*, 101 (1994), pp. 403–416.

L. Bankoff, The metamorphosis of the butterfly problem. *Mathematics Magazine*, 60 (1987), pp. 195–210.

A. Bell, Hansen's right triangle theorem, its converse and a generalization, *Forum Geometricorum*, 6 (2006), pp. 335–342.

A. T. Benjamin and J. J. Quinn, *Proofs That Really Count*, Mathematical Association of America, Washington, 2003.

M. Bicknell and V. E. Hoggatt Jr., eds., *A Primer for the Fibonacci Numbers*, The Fibonacci Association, San Jose, 1972.

E. M. Bishop, The use of the pentagram in constructing the net for a regular dodecahedron. *Mathematical Gazette*, 46 (1962), p. 307.

I. C. Bivens and B. G. Klein, Geometric series. *Mathematics Magazine*, 61 (1988), p. 219.

W. J. Blundon, On certain polynomial associated with the triangle, *Mathematics Magazine* 36 (1963), pp. 247–248.

A. Bogomolny, The Taylor circle, from *Interactive Mathematics Miscellany and Puzzles*, http://www.cut-the-knot.org/triangle/Taylor.shtml, accessed 29 April 2010.

———, Three circles and common chords from *Interactive Mathematics Miscellany and Puzzles*, http://www.cut-the-knot.org/proofs/circlesAndSpheres.shtml, accessed 19 February 2010.

B. Bolt, R. Eggleton, and J. Gilks, The magic hexagon. *Mathematical Gazette*, 75 (1991), pp. 140–142.

J. Bonet, *The Essential Gaudí. The Geometric Modulation of the Church of the Sagrada Familia*. Portic, Barcelona, 2000.

R. Bracho López, *El Gancho Matemático, Actividades Recreativas para el Aula*. Port-Royal Ediciones, Granada, 2000.

B. Bradie, Exact values for the sine and cosine of multiples of 18°. *College Mathematics Journal*, 33 (2002), pp. 318–319.

H. C. Bradley, Solution to problem 3028, *American Mathematical Monthly*, 37 (1930), pp. 158–159.

B. Branner, The Mandelbrot set, in R. L. Devaney and L. Keen (eds.), *Chaos and Fractals* (Proceedings of Symposia in Applied Mathematics, 39), American Mathematical Society, Providence, 1989, pp. 75–105.

A. Brousseau, Sums of squares of Fibonacci numbers, in *A Primer for the Fibonacci Numbers*, M. Bicknell and V. E. Hoggatt Jr., eds., The Fibonacci Association, San Jose, 1972, p. 147.

F. Burk, Behold: The Pythagorean theorem, *College Mathematics Journal*, 27 (1996), p. 409.

R. P. Burn, Gregory of St. Vincent and the rectangular hyperbola, *Mathematical Gazette*, 84 (2000), pp. 480–485.

S. Burr, Planting trees. In D. A. Klarner (ed.), *The Mathematical Gardner*. Prindle, Weber & Schmidt, Boston, 1981, pp. 90–99.

F. Cajori, Historical note, *American Mathematical Monthly*, 6 (1899), pp. 72–73.

B. C. Carlson, The logarithmic mean, *American Mathematical Monthly*, 79 (1972), pp. 615–618.

D. S. Clark and O. Shisha, Proof without words: Inductive construction of an infinite chessboard with maximal placement of nonattacking queens. *Mathematics Magazine*, 61 (1988), p. 98.

S. Coble, Proof without words: An efficient trisection of a line segment. *Mathematics Magazine*, 67 (1994), p. 354.

J. H. Conway, The power of mathematics, in A. F. Blackwell and D. J. C. MacKay, *Power*, Cambridge University Press, Cambridge, 2005, pp. 36–50.

J. H. Conway and R. Guy, *The Book of Numbers*. Copernicus, New York, 1996.

H. S. M. Coxeter and S. L. Greitzer, *Geometry Revisited*, Mathematical Association of America, Washington, 1967.

A. Cupillari, Proof without words, *Mathematics Magazine*, 62 (1989), p. 259.

W. Dancer, Geometric proofs of multiple angle formulas. *American Mathematical Monthly*, 44 (1937), pp. 366–367.

R. de la Hoz, La proporción cordobesa. *Actos VII Jornadas Andaluzas de Educación Matemática*, Publicaciones Thales, Córdoba (1995), pp. 65–74.

M. de Villiers, Stars: A second look. *Mathematics in School*, 28 (1999), p. 30.

D. DeTemple and S. Harold, A round-up of square problems, *Mathematics Magazine*, 69 (1996), pp. 15–27.

P. Deiermann, The method of last resort (Weierstrass substitution), *College Mathematics Journal*, 29 (1998), p. 17.

H. Dudeney, *The Canterbury Puzzles and Other Curious Problems*, T. Nelson and Sons, London, 1907 (reprinted by Dover Publications, Inc., New York, 1958).

———, *Amusements in Mathematics*. T. Nelson and Sons, London, 1917 (reprinted by Dover Publications, Inc., New York, 1958).

T. Duval, Preuve sans parole, *Tangente*, 115 (Mars-Avril 2007), p. 10.

R. H. Eddy, A theorem about right triangles, *College Mathematics Journal*, 22 (1991), p. 420.

———, Proof without words. *Mathematics Magazine*, 65 (1992), p. 356.

P. Erdős, Problem and Solution 3746. *American Mathematical Monthly*, 44 (1937), p. 400.

M. A. Esteban, *Problemas de Geometria*. Federación Española de Sociedades de Profesores de Matemáticas, Cáceres, Spain, 2004.

L. Euler, in: *Leonhard Euler und Christian Goldbach, Briefwechsel 1729–1764*, A.P. Juskevic and E. Winter (editors), Akademie Verlag, Berlin, 1965.

R. G. Everitt, Corollaries to the "chord and tangent" theorem. *Mathematical Gazette*, 34 (1950), pp. 200–201.

H. Eves, *In Mathematical Circles*. Prindle, Weber & Schmidt, Inc., Boston, 1969 (reprinted by Mathematical Association of America, Washington, 2003).

———, *Great Moments in Mathematics (Before* 1650), Mathematical Association of America, Washington, 1980.

A. Flores, Tiling with squares and parallelograms. *College Mathematics Journal*, 28 (1997), p. 171.

J. W. Freeman, The number of regions determined by a convex polygon, *Mathematics Magazine*, 49 (1976), pp. 23–25.

G. N. Frederickson, *Tilings, Plane & Fancy*, Cambridge University Press, New York, 1997.

K. Follett, *The Pillars of the Earth*. William Morrow and Company, Inc., New York, 1989.

V. W. Foss, Centre of gravity of a quadrilateral. *Mathematical Gazette*, 43 (1959), p. 46.

C. L. Frenzen, Proof without words: Sums of consecutive positive integers, *Mathematics Magazine*, 70 (1997), p. 294.

B. C. Gallivan, *How to Fold Paper in Half Twelve Times: An Impossible Challenge Solved and Explained*. Historical Society of Pomona Valley, Pomona CA (2002).

C. Gamer, D. W. Roeder, and J. J. Watkins, Trapezoidal numbers. *Mathematics Magazine*, 58 (1985), pp. 108–110.

M. Gardner, *Mathematical Carnival*, Alfred A. Knopf, Inc., New York, 1975.

————, Some new results on magic hexagrams, *College Mathematics Journal*, 31 (2000), pp. 274–280.

C. Gattegno, La enseñanza por el filme matemático, in C. Gattegno et al. eds, *El Material Para la Enseñanza de las Matemáticas, 2° ed.*, Editorial Aguilar, Madrid, 1967, pp. 97–111.

P. Glaister, Golden earrings. *Mathematical Gazette*, 80 (1996), pp. 224–225.

S. Golomb, A geometric proof of a famous identity, *Mathematical Gazette*, 49 (1965), pp. 198–200.

J. Gomez, Proof without words: Pythagorean triples and factorizations of even squares. *Mathematics Magazine*, 78 (2005), p. 14.

A. Gutierrez, *Geometry Step-by-Step from the Land of the Incas*. www.agutie.com accessed 21 June 2010.

D. W. Hansen, On inscribed and escribed circles of right triangles, circumscribed triangles, and the four square, three square problem, *Mathematics Teacher*, 96 (2003), 358–364.

T. L. Heath, *The Works of Archimedes*, Cambridge University Press, Cambridge, 1897.

V. E. Hill IV, President Garfield and the Pythagorean theorem, *Math Horizons*, February 2002, pp. 9–11, 15.

L. Hoehn, A simple generalisation of Ceva's theorem. *Mathematical Gazette*, 73 (1989), pp. 126–127.

————, Proof without words, *College Mathematics Journal*, 35 (2004), p. 282.

K. Hofstetter, A simple construction of the golden section. *Forum Geometricorum*, 2 (2002), pp. 65–66.

R. Honsberger, *Mathematical Gems*, Mathematical Association of America, Washington, 1973.

————, *Mathematical Gems II*. Mathematical Association of America, Washington, 1976.

————, *Mathematical Morsels*. Mathematical Association of America, Washington, 1978.

————, *Mathematical Gems III*, Mathematical Association of America, Washington, 1985.

————, *Episodes in Nineteenth and Twentieth Century Euclidean Geometry*. Mathematical Association of America, Washington, 1995.

————, *Mathematical Delights*. Mathematical Association of America, Washington, 1995.

————, *Mathematical Chestnuts from Around the World*. Mathematical Association of America, Washington, 2001.

D. Houston, Proof without words: Pythagorean triples via double angle formulas. *Mathematics Magazine*, 67 (1994), p. 187.

N. Hungerbühler, Proof without words: The triangle of medians has three-fourths the area of the original triangle. *Mathematics Magazine*, 72 (1999), p. 142.

R. A. Johnson, A circle theorem. *American Mathematical Monthly*, 23 (1916), pp. 161–162.

W. Johnston and J. Kennedy, Heptasection of a triangle. *Mathematics Teacher*, 86 (1993), p. 192.

N. D. Kazarinoff, *Geometric Inequalities*. Mathematical Association of America, Washington, 1961.

A. B. Kempe, *How to Draw a Straight Line: A Lecture on Linkages*. Macmillan and Company, London, 1877.

C. Kimberling, *Encyclopedia of Triangle Centers*, <http://faculty.evansville.edu/ck6/encyclopedia/ETC.html>, accessed 6 April 2010.

Y. Kobayashi, A geometric inequality, *Mathematical Gazette*, 86 (2002), p. 293.

J. D. E. Konhauser, D. Velleman, and S. Wagon, *Which Way Did the Bicycle Go?* Mathematical Association of America, Washington, 1996.

S. H. Kung, Proof without words: The Weierstrass substitution, *Mathematics Magazine*, 74 (2001), p. 393.

———, Proof without words: The Cauchy-Schwarz inequality, *Mathematics Magazine*, 81 (2008), p. 69.

G. Lamé, Un polygone convexe étant donné, de combien de manières peut-on le partager en triangles au moyen de diagonals? *Journal de Mathématiques Pures et Appliquées*, 3 (1838), pp. 505–507.

L. H. Lange, Several hyperbolic encounters. *Two-Year College Mathematics Journal*, 7 (1976), pp. 2–6.

L. Larson, A discrete look at $1 + 2 + \cdots + n$. *College Mathematics Journal*, 16 (1985), pp. 369–382.

C.-S. Lee, Polishing the star. *College Mathematics Journal*, 29 (1998), pp. 144–145.

B. Lindström and H.-O. Zetterström, Borromean circles are impossible. *American Mathematical Monthly*, 98 (1991), pp. 340–341.

C. T. Long, On the radii of the inscribed and escribed circles of right triangles—A second look, *Two-Year College Mathematics Journal*, 14 (1983), pp. 382–389.

E. S. Loomis, *The Pythagorean Proposition*, National Council of Teachers of Mathematics, Reston, VA, 1968.

S. Loyd, *Sam Loyd's Cyclopedia of 5000 Puzzles, Tricks, and Conundrums (With Answers)*, The Lamb Publishing Co., New York, 1914. Available online at http://www.mathpuzzle.com/loyd/

W. Lushbaugh, cited in S. Golomb, A geometric proof of a famous identity, *Mathematical Gazette*, 49 (1965), pp. 198–200.

R. Mabry, Proof without words. *Mathematics Magazine*, 72 (1999), p. 63.

————, Mathematics without words. *College Mathematics Journal*, 32 (2001), p. 19.

B. Mandelbrot, *Fractals: Form, Chance, and Dimension*. W. H. Freeman, San Francisco, 1977.

G. E. Martin, *Polyominoes: A Guide to Puzzles and Problems in Tiling*. Mathematical Association of America, Washington, 1991.

M. Moran Cabre, Mathematics without words. *College Mathematics Journal*, 34 (2003), p. 172.

F. Nakhli, Behold! The vertex angles of a star sum to 180°. *College Mathematics Journal*, 17 (1986), p. 338.

R. B. Nelsen, Proof without words: The area of a salinon, *Mathematics Magazine*, 75 (2002a), p. 130.

————, Proof without words: The area of an arbelos, *Mathematics Magazine*, 75 (2002b), p. 144.

————, Proof without words: Lunes and the regular hexagon. *Mathematics Magazine*, 75 (2002c), p. 316.

————, Mathematics without words: Another Pythagorean-like theorem. *College Mathematics Journal*, 35 (2004), p. 215.

————, Proof without words: A triangular sum. *Mathematics Magazine*, 78 (2005), p. 395.

————, Proof without words: Inclusion-exclusion for triangular numbers. *Mathematics Magazine*, 79 (2006), p. 65.

I. Niven, *Maxima and Minima Without Calculus*, Mathematical Association of America, Washington, 1981.

S. Okuda, Proof without words: The triple-angle formulas for sine and cosine. *Mathematics Magazine*, 74 (2001), p. 135.

R. L. Ollerton, Proof without words: Fibonacci tiles, *Mathematics Magazine*, 81 (2008), p. 302.

H. Ouellette and G. Bennett, The discovery of a generalization: An example in problem-solving. *Two-Year College Mathematics Journal*, 10 (1979), pp. 100–106.

D. Pedoe, *Geometry and the Liberal Arts*. St. Martin's Press, New York, 1976.

Á. Plaza, Proof without words: Mengoli's series. *Mathematics Magazine*, 83 (2010), p. 140.

A. B. Powell, Caleb Gattegno (1911–1988): A famous mathematics educator from Africa? *Revista Brasiliera de História de Matematica*, Especial no. 1 (2007), pp. 199–209.

R. Pratt, Proof without words: A tangent inequality. *Mathematics Magazine*, 83 (2010), p. 110.

V. Priebe and E. A. Ramos, Proof without words: The sine of a sum, *Mathematics Magazine*, 73 (2000), p. 392.

A. D. Rawlins, A note on the golden ratio. *Mathematical Gazette*, 79 (1995), p. 104.

J. F. Rigby, Napoleon, Escher, and tessellations. *Mathematics Magazine*, 64 (1991), pp. 242–246.

P. L. Rosin, On Serlio's construction of ovals. *Mathematical Intelligencer*, 23:1 (Winter 2001), pp. 58–69.

J. C. Salazar, Fuss' theorem. *Mathematical Gazette*, 90 (2006), pp. 306–307.

J. Satterly, The nedians of a triangle. *Mathematical Gazette*, 38 (1954), pp. 111–113.

———, The nedians, the nedian triangle and the aliquot triangle of a plane triangle. *Mathematical Gazette*, 40 (1956), pp. 109–113.

D. Schattschneider, Proof without words: The arithmetic mean-geometric mean inequality, *Mathematics Magazine*, 59 (1986), p. 11.

N. Schaumberger, An alternate classroom proof of the familiar limit for e, *Two-Year College Mathematics Journal*, 3 (1972), pp. 72–73.

K. Scherer, Difficult dissections. http://karl.kiwi.gen.nz/prdiss.html, accessed 2 April 2010.

D. B. Sher, Sums of powers of three. *Mathematics and Computer Education*, 31:2 (Spring 1997), p. 190.

W. Sierpiński, *Pythagorean Triangles*. Yeshiva University, New York, 1962.

T. A. Sipka, The law of cosines, *Mathematics Magazine*, 61 (1988), p. 113.

S. L. Snover, Four triangles with equal area, in: R. Nelsen, *Proofs Without Words II*, Mathematical Association of America, Washington, 2000, p. 15.

S. L. Snover, C. Waivaris, and J. K. Williams, Rep-tiling for triangles. *Discrete Mathematics*, 91 (1991), pp. 193–200.

J. Struther, *Mrs. Miniver*, Harcourt Brace Jovanovich, Publishers, San Diego, 1990.

R. Styer, Trisecting a line segment. http://www41.homepage.villanova.edu/robert.styer/trisecting segment/. Page dated 25 April 2001.

J. H. Tanner and J. Allen, *An Elementary Course in Analytic Geometry*. American Book Company, New York, 1898.

J. Tanton, Proof without words: Geometric series formula. *College Mathematics Journal*, 39 (2008), p. 106.

————, Proof without words: Powers of two. *College Mathematics Journal*, 40 (2009), p. 86.

M. G. Teigen and D. W. Hadwin, On generating Pythagorean triples. *American Mathematical Monthly*, 78 (1971), pp. 378–379.

C. W. Trigg, Bisection of yin and of yang, *Mathematics Magazine*, 34 (1960), pp. 107–108.

F. van Lamoen, Friendship among triangle centers, *Forum Geometricorum*, 1 (2001), pp. 1–6.

J. Venn, On the diagrammatic and mechanical representation of propositions and reasonings. *The London, Edinburgh, and Dublin Philosophical Magazine and Journal of Science*, 9 (1880), pp. 1–18.

D. B. Wagner, A proof of the Pythagorean theorem by Liu Hui (third century A.D.), *Historia Mathematica*, 12 (1985), pp. 71–73.

S. Wagon, Fourteen proofs of a result about tiling a rectangle. *American Mathematical Monthly*, 94 (1987), pp. 601–617.

S. Wakin, Proof without words: 1 domino = 2 squares: Concentric squares, *Mathematics Magazine*, 60 (1987), p. 327.

R. J. Walker, Note by the editor, *American Mathematical Monthly*, 49 (1942), p. 325.

I. Warburton, Bride's chair revisited again!, *Mathematical Gazette*, 80 (1996), pp. 557–558.

W. T. Webber and M. Bode, Proof without words: The cosine of a difference, *Mathematics Magazine*, 75 (2002), p. 398.

D. Wells, *The Penguin Dictionary of Curious and Interesting Geometry*. Penguin Books, London, 1991.

R. S. Williamson, A formula for rational right-angles triangles, *Mathematical Gazette*, 37 (1953), pp. 289–290.

R. Woods, The trigonometric functions of half or double an angle, *American Mathematical Monthly*, 43 (1936), pp. 174–175.

R. H. Wu, Arctangent identities, *College Mathematics Journal*, 34 (2003), pp. 115, 138.

———, Euler's arctangent identity, *Mathematics Magazine*, 77 (2004), pp. 189.

K. L. Yocum, Square in a Pythagorean triangle. *College Mathematics Journal*, 21 (1990), pp. 154–155.

Index

About the Authors

Claudi Alsina was born on 30 January 1952 in Barcelona, Spain. He received his BA and PhD in mathematics from the University of Barcelona. His post-doctoral studies were at the University of Massachusetts, Amherst, Claudi, Professor of Mathematics at the Technical University of Catalonia has developed a wide range of international activities, research papers, publications and hundreds of lectures on mathematics and mathematics education. His latest books include *Associative Functions: Triangular Norms and Copulas* (with M.J. Frank and B. Schweizer) WSP, 2006; *Math Made Visual. Creating Images for Understanding Mathematics* (with Roger Nelsen) MAA, 2006; *Vitaminas Matemáticas* and *El Club de la Hipotenusa*, Ariel, 2008, *Geometria para Turistas*, Ariel, 2009, *When Less Is More: Visualizing Basic Inequalities* (with Roger Nelsen) MAA, 2009; *Asesinatos Matemáticos*, Ariel, 2010 and *Charming Proofs: A Journey Into Elegant Mathematics* (with Roger Nelsen) MAA, 2010.

Roger B. Nelsen was born on 20 December 1942 in Chicago, Illinois. He received his B.A. in mathematics from DePauw University in 1964 and his Ph.D. in mathematics from Duke University in 1969. Roger was elected to Phi Beta Kappa and Sigma Xi, and taught mathematics and statistics at Lewis & Clark College for forty years before his retirement in 2009. His previous books include *Proofs Without Words: Exercises in Visual Thinking*, MAA 1993; *An Introduction to Copulas*, Springer, 1999 (2nd. ed. 2006); *Proofs Without Words II: More Exercises in Visual Thinking*, MAA, 2000; *Math Made Visual: Creating Images for Understanding Mathematics* (with Claudi Alsina), MAA, 2006; *When Less Is More: Visualizing Basic Inequalities* (with Claudi Alsina), MAA, 2009; *Charming Proofs: A Journey Into Elegant Mathematics* (with Claudi Alsina), MAA, 2010; and *The Calculus Collection: A Resource for AP and Beyond* (with Caren Diefenderfer, editors), MAA, 2010.